T0320599

LOGIC MINIMIZATION ALGORITHMS FOR VLSI SYNTHESIS

THE KLUWER INTERNATIONAL SERIES IN ENGINEERING AND COMPUTER SCIENCE

VLSI, Computer Architecture, and Digital Signal Processing

Consulting Editor: Jonathan Allen

LOGIC MINIMIZATION ALGORITHMS FOR VLSI SYNTHESIS

by

ROBERT K. BRAYTON
GARY D. HACHTEL
CURTIS T. McMULLEN
ALBERTO L. SANGIOVANNI-VINCENTELLI

KLUWER ACADEMIC PUBLISHERS
Boston/The Hague/Dordrecht/Lancaster

Distributors for North America:
KLUWER ACADEMIC PUBLISHERS
190 Old Derby Street
Hingham, MA 02043, U.S.A.

Distributors outside North America:
KLUWER ACADEMIC PUBLISHERS GROUP
Distribution Centre
P.O. Box 322
3300 AH Dordrecht
THE NETHERLANDS

Library of Congress Cataloging in Publication Data

Main entry under title:

Logic minimization algorithms for VLSI synthesis.

(The Kluwer international series in engineering and computer science)
Bibliography: p. 174
Includes index.
1. Logic design. 2. Integrated circuits--Very large scale integration. 3. Integrated circuits--Design and construction--Data processing. 4. Algorithms. I. Brayton, Robert King. II. Title. III. Series.
TK7868.L6E78 1984 621.381'73 84-15490
ISBN 0-89838-164-9

Eighth Printing 1997

Printed in the United States of America

Table of Contents

Table of Contents

Preface

The roots of the project which culminates with the writing of this book can be traced to the work on logic synthesis started in 1979 at the IBM Watson Research Center and at University of California, Berkeley. During the preliminary phases of these projects, the importance of logic minimization for the synthesis of area and performance effective circuits clearly emerged. In 1980, Richard Newton stirred our interest by pointing out new heuristic algorithms for two-level logic minimization and the potential for improving upon existing approaches.

In the summer of 1981, the authors organized and participated in a seminar on logic manipulation at IBM Research. One of the goals of the seminar was to study the literature on logic minimization and to look at heuristic algorithms from a fundamental and comparative point of view. The fruits of this investigation were surprisingly abundant: it was apparent from an initial implementation of recursive logic minimization (ESPRESSO-I) that, if we merged our new results into a two-level minimization program, an important step forward in automatic logic synthesis could result. ESPRESSO-II was born and an APL implementation was created in the summer of 1982. The results of preliminary tests on a fairly large set of industrial examples were good enough to justify the publication of our algorithms. It is hoped that the strength and speed of our minimizer warrant its Italian name, which denotes both express delivery and a specially-brewed black coffee.

We decided to publish this material as a monograph rather than as a series of papers for several reasons. We wanted to maintain a unified point of view and develop the necessary foundational material; to provide a reference for the mathematical results on which the procedures rest; and finally to describe in depth the algorithms themselves.

To describe the algorithms, we have used a pseudo-computer language. Our goal was to enable a competent reader to implement the algorithms directly. We feel that this goal has been achieved although many of the details of implementation are missing. In the fall of 1983, Richard Rudell, a graduate student at Berkeley, working from an early rough draft of this book, implemented ESPRESSO-II in the C language. An initial working version was available in January 1984. It is clear from our experiments, comparisons, and discussions with Rudell that ESPRESSO-IIC is a complete and high quality implementation of the ESPRESSO-II algorithms. Further refinements planned by Rudell are being implemented currently.

ESPRESSO-IIC is available, for nominal tape handling charges, from the Industrial Liaison Program, Department of EECS, University of California, Berkeley, CA 94720. It exploits certain features of the C language to provide an efficient machine independent implementation. In fact, ESPRESSO-IIC has been run on the DEC VAX 11/780 under Berkeley 4.2 Unix, the IBM 3081 under VM/CMS, the SUN and APOLLO 68000 based workstations and the IBM 8088 based personal computer. Only a small conversion effort was required to transport ESPRESSO-IIC to these dissimilar machines.

The simultaneous availability of a program and a book providing a complete description of its algorithms and their mathematical justification is unusual. This approach fits well into the scope of the series edited by J. Allen. We hope that it will set a precedent for similar offerings in the CAD area.

A rather large set of references on logic minimization and related topics has been accumulated in the course of writing this book. We make them available in the book even though many of them are not directly referenced. The references are not intended to be a complete bibliography in this field and we apologize in advance to those authors whose contributions have not been noted.

This book is not the final word on logic minimization nor is it intended to be. We hope that by providing a complete and rigorous description of the ESPRESSO algorithms and a computer implementation, the readers will be able to experiment with variations on the overall iterative scheme and different heuristics in the algorithms themselves. To aid experimentation in this area, the PLA's which were the basis for our experiments are included with ESPRESSO-IIC. These PLA's form a useful set of test cases since they represent a good mixture of "realistic" problems and known difficult examples.

We also hope that combinations of some of our algorithms, possibly restricted to the single-output case, will find their way into multi-level or multi-stage logic synthesis programs. Perhaps this book

book will provide some basis for future development in this relatively new area.

We wish to thank Richard Rudell for his feedback from several readings of this manuscript and for many good ideas which became part of ESPRESSO-II. His excellent implementation of ESPRESSO-IIC led to improved algorithms and better heuristics. Secondly, we thank Barbara White for her tireless efforts in typing and formatting this manuscript to provide the publisher with camera-ready copy. Initial versions of the manuscript were aided by Jo Genzano and Ruth Major. Many colleagues interacted with us over the years and we want to thank Karen Bartlett, Josh Cohen, Giovanni DeMicheli, E. Eschen, Lane Hemachandra, Se June Hong, Mike Lightner, Richard Newton, Tsutomo Sasao, Barry Trager, and David Yun. A special thanks to Dave DeWitt (posthumously) and Bob Risch for sparking the chain of ideas leading from Cascode Current Switch to these efforts.

Financial assistance for this project was provided by IBM, the National Science Foundation under grant ECS-8340435, and the Defense Advanced Research Project Agency under grant N-00039-84-0107. Fairchild, GTE, Harris Semiconductors and the MICRO program of the State of California are acknowledged for their assistance in the ESPRESSO-IIC project.

Finally we thank our families for their kind support during the years devoted to this effort, and give a special thank you to "our Italian wife" whose pizza and espresso were an inspiration for this book.

LOGIC MINIMIZATION ALGORITHMS FOR VLSI SYNTHESIS

Chapter 1

INTRODUCTION

Very Large Scale Integration (VLSI) plays a major role in the development of complex electronic systems. As a result of the continued progress in semiconductor technology, an increasing number of devices can be put on a single micro-electronic chip. Newly developed design methods for VLSI must be able to cope with this increased design complexity [MEA 80]. Many such methods are required for the final implementation of a chip. This book addresses one area, that of logic minimization, and provides a new set of optimization algorithms which can cope with the increased complexity of VLSI designs.

The design of a digital system can be viewed as a sequence of transformations of design representations at different levels of abstraction. The behavioral specifications of the system are described first by a **functional representation.** For example, Hardware Description Languages such as ISP [BAR 77], [BAR 81], IDL [MAI 82] and AHPL [HIL 78], are used. The functional representation is transformed into a **logic description,** consisting of a net-list of logic gates, including storage gates, that can be visualized by a logic schematic diagram. Subsequently **logic synthesis** is performed, which transforms and minimizes the representation, achieving a technology-independent set of logical expressions. The next step is to specify an **electrical representation,** according to an implementation technology, which is eventually transformed into a **geometric layout** of the integrated circuit implementing the given functionality.

In an automated synthesis environment, the designer just monitors the transformations and verifies the practical feasibility of the design at each step. As a result, the length of the design cycle is shortened, the associated development cost is reduced and the circuit reliability is enhanced greatly.

1.1 Design Styles for VLSI Systems

The transformations involved in the synthesis of a VLSI chip depend on the design method used. Several "styles" are found in industrial VLSI design, depending on the function and the market of the product.

The objective of a fully **custom** design method is an *ad hoc* implementation. Each transformation between design representations is optimized for the particular application. High performance implementations can be achieved. To date, computer-aided design techniques support custom design to a limited extent, but it is not yet known how to replace designers' experience by automated transformations between design representations. As a consequence, custom design has a longer development time compared to other methods. Thus custom design is profitable to date only for large volume production of complex systems, such as microprocessors or memories, or for circuits where special performance is required.

In a **gate-array** design method, a circuit is implemented in silicon by personalizing a master array of uncommitted gates using a set of interconnections [BLU 79]. The design is constrained by the fixed structure of the master array and is limited to routing the interconnections. Computer-aids for gate-array design allow complex circuits to be implemented in a short time. Gate-arrays are widely used, in particular for small volume production or for prototyping new designs.

The design of a VLSI circuit in a **standard-cell** (or **poly-cell**) design method requires partitioning the circuit into atomic units that are implemented by precommitted cells (e.g. [FEL 76]). Placement and routing of the cells is supported by computer-aided design tools (e.g. [PER 77]). The standard-cell and gate-array design methods alone do not support highly optimized designs. Standard cell designs are more flexible than gate-array designs, but require longer development time.

The design of VLSI circuits, using algorithmically generated **macro-cells,** bridges the gap between custom and standard-cell design and is compatible with both methods. Macro-cells can implement functional units that are specified by design parameters and by their functionality [NEW 81]. Macro-cells are usually highly regular and structured allowing computer programs, called **module generators,** to produce the layout of a macro-cell from its functional description.

The macro-cell approach is very attractive because its flexibility allows the designer to exploit the advantages of both custom and standard-cell methods. Highly optimized and area-efficient modules can be designed in a short time. In particular, **Programmable Logic Array** (PLA) macros have been shown to be very effective for designing both combinatorial and sequential functions [FLE 75], [PRO 76].

Weinberger proposed a regular one-dimensional array implementation of a set of logic functions [WEI 67]. A structured approach for the implementation of combinational logic functions can also be obtained by look-up tables [FLE 75]. In particular **Read Only Memories** (ROM) can be used to evaluate a logic function whose entries are stored in the memory locations corresponding to appropriate addresses. Read Only Memories are two-dimensional arrays and are implemented in several technologies. The implementation of a combinational logic function by means of a ROM yields a very structured design, which can be easily modified. Moreover, the speed of operation can be fast and easily determined for the set of combinational logic functions requiring the same address space. Unfortunately a ROM implementation of a combinational function is not effective in terms of silicon area. A ROM implementation of a n-input function requires 2^n memory cells. In addition, a combinational function may not require the entire address space of a ROM because the function is not specified for some input combinations.

Programmable Logic Arrays allow for more efficient use of silicon area by representing a set of logic functions in a compressed yet regular form. A PLA implements two-stage combinational logic (e.g. OR of ANDs) through an adjacent pair of rectangular arrays.

Each position of the array is programmed by the presence or absence of a device. The functionality of a PLA can be represented by a 0-1 matrix, and the design and optimization of a PLA can be carried out directly in terms of this functional description. PLAs are attractive building blocks of a structured design methodology [LOG 75]. Many industrial VLSI circuits, such as the Intel 8086, the Motorola 68000, the Hewlett-Packard 32-bit and the Bell Mac-32 micro-processors, use PLAs as building blocks.

Sequential logic functions can be represented as **Finite State Machines** (FSMs) and implemented by a combinational and a storage component [HIL 81]. A regular array can implement effectively the FSM combinational component. In particular, PLA-based Finite State Machines can be designed efficiently, because the properties of two-level combinational functions are well understood. PLAs and memory elements can be seen as primitives of a general digital design methodology.

1.2 Automatic Logic Synthesis

The automated synthesis of a combinational system can be partitioned into several tasks: **functional design, logic design,** and **physical design.**

Functional design consists of defining the functional specification of the system. For example, a designer can specify a combinational circuit in a Hardware Description Language. The functional description is then transformed into a logic description in terms of Boolean (logic) variables.

Logic design, the major focus of this book, consists of a manipulation of the logic representation without modifying the functionality. Optimal logic design transforms the logic representation to obtain a convenient implementation. In particular **logic minimization,** [MCK 56],

[HON 74], [BRO 81], [BRA 82a], and **logic partitioning** [BRA 82b] have been explored. Two-level logic minimization seeks a logic representation with a minimal number of implicants and literals. Reduction of the number of implicants allows a PLA to be implemented in a smaller area, while reduction of the number of literals corresponds to a reduction in the number of devices and contacts required. Both lead to a greater speed of operation, higher yield and enhanced reliability.

Physical design involves the placement and interconnection of the devices. For a PLA, it is possible to define a straight-forward mapping from a two-level logic description to the PLA layout. Optimal physical design increases the area utilization by rearranging the array structure. For PLA's this can be done by **array folding** and **array partitioning.** (See [HAC 82], [DEM 83c], [DEM 83e], [HEN 83].)

PLA design depends heavily on the integrated circuit design method. PLAs that are built in gate-array chips have to be designed so that they fit into a given structure. Therefore physical optimization is very important in obtaining compact arrays with a given shape. On the other hand, in VLSI custom and macro-cell design, the PLA shape is not the major design constraint. PLAs must interact with other functional building-blocks, and a primary objective of design is to ease the connection of the PLA to other subcircuits.

The final representation is a layout of the masks used to manufacture the circuit. This step depends heavily on the semiconductor process and layout design rules and is described using a geometric design language such as the *Caltech Intermediate Form* (CIF) [MEA 80] or equivalent mask language.

1.3 PLA Implementation

A combinational logic function can be described by a logic cover. When designing a PLA implementation of a combinational function, the logic cover is represented by a pair of matrices, called **input** and **output** matrices.

The input and output matrices shown on the left and right below

```
**1**0 1000
*1*0** 0100
1****0 0001
1***1* 0100
0***** 0010
*****1 0001
```

have a corresponding PLA implementation shown in Figure 1.1

Every input to the logic function (each column of the input matrix) corresponds to a pair of columns in the left part of the physical array. Every implicant, or equivalently every row of the input and output matrices, corresponds to a row of the physical array. Every input row corresponds to a logical product of some set of inputs. Every output of the logic function (each column of the output matrix) corresponds to a column in the right part of the physical array. The implementation of a particular switching function is obtained by "programming" the PLA, i.e. by placing (or connecting) appropriate devices in the array in the input or output column position specified by "1" or "0". The input and output arrays are referred to as the **AND-plane** and **OR-plane** respectively, although physical implementations different from *sum-of-products* (OR of ANDs) are very common.

The key technological advantage of using a PLA in an integrated circuit technology relies on the straight-forward mapping between the symbolic representation and its physical implementation. Moreover PLAs are compatible with different technologies and modes of operations, such as **AND-OR bipolar transistor PLA** implementations and **NOR-NOR NMOS PLA** implementations [DEM 84]. The **NOR-NOR** implementation is most common in VLSI MOS circuits [MEA 80] since in MOS technology it is convenient to exploit the use of NOR gates. PLAs are implemented in the form of *sum-of-sums* (or more exactly *complemented-sum-of-complemented-sums).* The transformation from a *sum-of-products* representation can be easily carried out. Note that output inverters are required.

The basic NMOS PLA implementation has been tailored to different MOS processes yielding better performance. High perform-

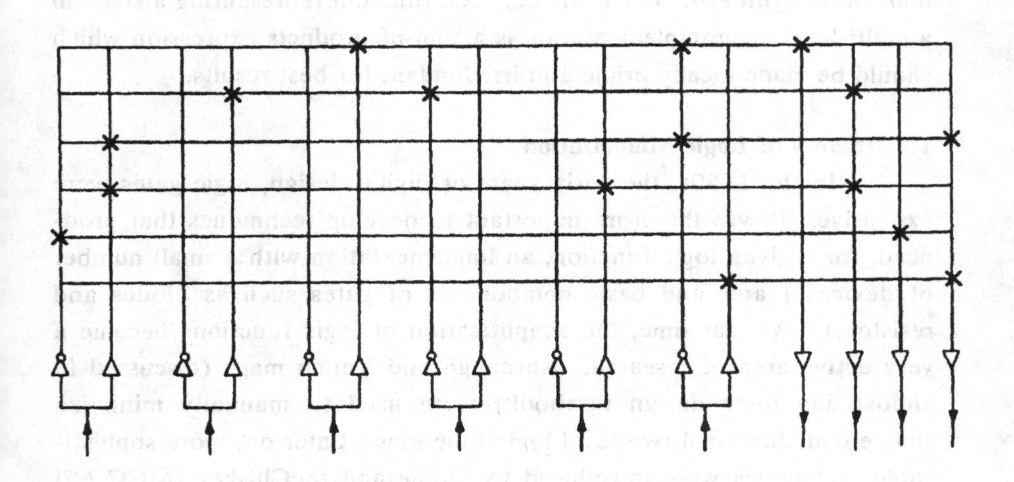

Figure 1.1: PLA physical implementation

ance NMOS implementations with static and dynamic operation are presented in [COO 79] and dynamic CMOS implementations have been obtained using the same basic structure. It is worth mentioning that up until the definition of the device and interconnection locations, PLA design is independent of the implementation technology. Therefore methods for PLA automated synthesis at the functional, logical, and optimization levels are fairly general and have a wide range of applications.

The major focus of this book is the area of logic minimization and the presentation of new algorithms appropriate for VLSI applications. Although much of the discussion concerns logic minimization in a setting appropriate for PLA implementation, i.e., the minimization of two-level, multiple-output logic functions, the application of these algorithms to logic synthesis in general should not be overlooked. While PLA synthesis is the most mature application of logic minimization, the area of logic synthesis of multi-level logic is very active. Certainly, two-level single-output logic minimization is an important ingredient of multi-level synthesis. For example, each function representing a stage in a multi-level logic implementation is a sum-of-products expression which should be made locally prime and irredundant for best results.

1.4 History of Logic Minimization

In the 1950s, the early years of digital design, logic gates were expensive. It was therefore important to develop techniques that produced, for a given logic function, an implementation with a small number of devices (gates and basic components of gates such as diodes and resistors). At that time, the simplification of logic functions became a very active area of research. Karnaugh and Veitch maps (discussed in almost any logic design textbook) were used to manually minimize simple combinational two-level logic functions. Later on, more sophisticated techniques were introduced by Quine and McCluskey [MCC 65] to obtain a two-level implementation of a given combinational logic function with the minimum number of gates. These techniques involved two steps:

1. Generation of all prime implicants,
2. Extraction of a minimum prime cover.

Even though the generation of all prime implicants has become more efficient (see the program MINI developed by Hong, Cain and Ostapko at IBM) [HON 74], it can be shown that the number of prime implicants of a logic function with n input variables can be as large as $3^n/n$. In addition, the second step, in general implemented with a branch-and-bound technique, involves the solution of a minimum covering problem that is known to belong to the class of NP-complete problems. This leaves little hope of finding an efficient exact algorithm, i.e., a minimum covering algorithm whose running time is bounded by a polynomial in the number of elements in the covering problem. Since the number of elements in the covering problem may be proportional to the exponential of the number of input variables of the logic function, the use of these techniques is totally impractical even for medium sized problems (10-15 variables).

In the late 1960s and early 1970s, the cost of logic gates was reduced so that logic minimization was not as essential as before and hand-simplification was enough to produce satisfactory designs. In the late 1970s, the advent of LSI made regular structures such as PLAs very desirable for the implementation of logic functions, because of the reduction in the design time they offered.

Two-level logic minimization has again become an important tool in this design environment. The minimization of product terms obtained by logic minimization has a direct impact on the physical area required since each product term is implemented as a row of the PLA. Logic design for VLSI often involves logic functions with more than 30 inputs and outputs and with more than 100 product terms, for which exact minimization is impractical. The need for optimization in such cases has motivated various heuristic approaches to the problem.

One approach follows the structure of the classical logic minimization techniques, first generating all prime implicants. However, instead of generating a minimum cover, a near minimum cover is selected heuristically. This procedure still suffers from the possibility of generating a very large number of prime implicants. A second approach tries to simultaneously identify and select implicants (not necessarily prime) for the cover. Several algorithms have been proposed in this group (see [ROT 80], [HON 74], [RHY 77], [ARE 78], [BRO 81]). We review briefly some of the most significant results.

In two methods ([RHY 77], [ARE 78]) a base minterm of the care-set of the logic function to be minimized is selected. It is expanded until it is a prime and all minterms that are covered by this prime are removed. The procedure is iterated until all the minterms of the care-set are covered. In [RHY 77] all prime implicants containing the selected base minterm are generated and a prime that is either essential (i.e., that must be in any prime cover of the care-set of the logic function) or satisfies some heuristic criterion is selected. This method can be very inefficient due to the large number of possible prime implicants. In [ARE 78] only a subset of all prime implicants covering the base minterms is generated. This provides a much faster method, but the results are not as good as the first method. Although both these approaches have been implemented, they have been tested only on small logic functions (number of inputs less than 10).

More recently, heuristic approaches have found wide application in the design of practical PLAs. The earliest and most successful of these led to the program MINI developed at IBM in the middle 70s [HON 74]. Later a heuristic minimization program called PRESTO was introduced by D. Brown [BRO 81]. The ideas of Swoboda [SWO 79] have been credited as instrumental in this development [BRO 81]. Recently a version of MINI, called SPAM, containing partitioning to allow minimization of large PLAs, was implemented by Kang [KAN 81]. These approaches are centered around the procedure of expanding each implicant of the logic function and removing other implicants covered by the expanded implicant. MINI expands each implicant to its maximum size both in the input and output part. PRESTO expands the input part of each implicant to its maximum size, but then reduces the output parts of the implicants maximally by removing the implicants from as many output spaces as possible. The covering step is implicit in the output reduction step. In fact, an implicant is redundant and hence removed if its output part has been reduced to empty. In these approaches *implicants* are expanded and covered, thus eliminating a basic problem with the techniques previously mentioned: the generation and storage of all minterms. A program called SHRINK, [ROT 80] is a heuristic minimizer contained in the logic minimization program MIN370. MIN370 also contains an algorithm for generating all primes efficiently, based on the disjoint sharp operation. This is followed by the generation of an irredundant cover (i.e., a cover such that no proper subset is also a cover of the given logic function). The procedure of expansion and elimination of implicants covered by the expanded implicant produces a cover made of prime implicants but not necessarily an irredundant one. PRESTO's output reduction step guarantees the final cover is irredundant but not necessarily prime, since after reduction the implicants are not necessarily prime.

All of these basic strategies lead to local minima which may be far from the global minimum. To avoid this problem, after the first expansion and removal of covered implicants, MINI **reduces** the remaining implicants to their smallest size while still maintaining coverage of the logic function. The implicants are then examined in pairs to reshape them, enlarging one and reducing the other by the same set of minterms. Then the expansion process is repeated and the entire procedure is iterated until no reduction in the number of product terms is obtained. This strategy makes the exploration of larger regions in the optimization space possible, striving for better results at the expense of more computation time.

PRESTO differs from MINI in the way the expansion process is carried out. MINI generates the complement of the logic function to check if the expansion of an implicant does not change the coverage of the function. PRESTO avoids the initial cost of computing the complement, but then the input expansion process requires a check on whether all the minterms covered by the expanded implicant are covered by some other implicant of the cover, which, in general, costs more computation time.

MINI can minimize logic functions whose inputs are generated by two-bit decoders as proposed by Fleisher and Meissel [FLE 75]. On the average, PLAs with two-bit decoders require 10-20% smaller arrays than "standard" PLAs to realize the same logic functions. Thus, minimization of these arrays is of practical importance. This kind of logic minimization can be seen as a special case of multiple-value logic minimization. The IBM version of MINI performs general multiple-valued logic minimization.

During the summer of 1981 the authors created a program ESPRESSO-I [BRA 82a] to compare the various strategies employed by MINI and PRESTO in an environment where the efficiency of the implementations and the quality and sparsity required of the final result were controlled. ESPRESSO-I was a single program with many switches for controlling the sequence of actions. At the same time, experimentation with logic manipulation allowed us to improve some of the algorithms used in both the PRESTO approach and the MINI approach.

We concluded that the technique of computing the complement of the PLA was superior since on the average the initial cost involved (reduced by our discovery of better complementation algorithms [BRA 81c]) was offset by the more efficient expansion procedure that this allowed. Our evaluation took into account the speed and efficiency of the new fast complementation and tautology checking. We also felt that iteration as used by MINI is generally worthwhile for VLSI designs; the improvement in the minimization justifies the expenditure of additional computation time. The initial investment in computing the complement is then amortized over the additional steps required by the iterations.

1.5 ESPRESSO-II

During the summer of 1982, we began the task of creating a set of algorithms for logic minimization based on the previous year's comparisons and improved algorithms. The result is ESPRESSO-II, which basically follows the sequence of top-level transformations of iterated expansion-reduction pioneered by MINI.

Our goals in the design of ESPRESSO-II were to build a logic minimization tool such that in most cases

1 - the problems submitted by a logic designer could be solved with the use of limited computing resources;

2 - the final results would be close to a global optimum.

Since exact minimization is impractical, we depend upon MINI-style iterative improvement to achieve confidence in the optimality of the final result. Robustness is a result of efficient heuristic algorithms, as well as special case handling and dynamic partitioning (output splitting). Since many of the algorithms employed have worst-case exponential complexity, their success in practice ultimately rests upon the exploitation of special properties of the logic functions encountered in actual applications.

The sequence of operations carried out by ESPRESSO-II is outlined below.

ESPRESSO-II:

1. Complement (Compute the complement of the PLA and the don't-care set, i.e., compute the off-set).

2. Expand (Expand each implicant into a prime and remove covered implicants).
3. Essential Primes (Extract the essential primes and put them in the don't-care set).
4. Irredundant Cover (Find a minimal (optionally minimum) irredundant cover).
5. Reduce (Reduce each implicant to a minimum essential implicant).
6. Iterate 2, 4, and 5 until no improvement.
7. Lastgasp (Try reduce, expand and irredundant cover one last time using a different strategy. If successful, continue the iteration.)
8. Makesparse (Include the essential primes back into the cover and make the PLA structure as sparse as possible).

Although most of these procedures have a counterpart in MINI, the algorithms employed in ESPRESSO-II are new and quite different. Efficient Boolean manipulation is achieved through the "unate recursive paradigm", which is employed in complementation, tautology and other algorithms. In the expansion step, we introduce the notion of blocking and covering matrices, which allow for efficient and strategic selection of prime implicants. Irredundant cover uses a logic expression which gives the condition under which a given subset of cubes forms a cover, facilitating the intelligent choice of a minimal set. (MINI proceeds directly to Reduce without this operation, since a reduced cover is automatically irredundant.) Finally, the procedure LASTGASP allows us to guarantee at least a weak form of optimality for the final result: no single prime can be added to cover in such a way that two primes can then be eliminated.

1.6 Organization of the Book

The algorithms of ESPRESSO-II achieve a high level of efficiency, mainly through the use of a common theme, the "unate recursive paradigm". Roughly, following this paradigm a logic function is recursively divided until each part becomes unate. The required operation (e.g., complement, reduction, irredundant cover) is then performed on the unate function, usually in a highly efficient way. Finally, the results are merged appropriately to obtain the required result.

Chapters 2 and 3 provide the notation and theoretical foundations for logic and logic function manipulation. In Chapter 3 we introduce unate functions and basic results concerning them. The unate recursive paradigm is presented and illustrated with the SIMPLIFY algorithm. Although not part of ESPRESSO-II, SIMPLIFY provides a very fast heuristic method of logic minimization and is effective for single-output functions in applications where computation time is more important.

Chapter 4 is the technical heart of the book. Each of the algorithms of ESPRESSO-II is described in detail along with supporting theorems and proofs. In addition, the algorithms are outlined as structured procedures which can be found in the figures throughout the chapter.

As noted above, MINI allows bit-pairing and, more generally, multiple-valued inputs. In Chapter 5 we present a new method for using any two-valued logic minimizer to minimize logic functions with multiple-valued inputs. The method is very simple: it involves a particular encoding of the multiple-valued inputs and the creation of a don't-care set before the use of the two-valued logic minimizer. Our method is efficient and practical, so long as the number of possible values for a given input is not very large.

ESPRESSO-II has been implemented in APL and C, and tested on many PLAs. We were interested in the final size of the PLA, its sparsity, and the amount of computation time expended in achieving the result. The PLAs tested included control logic as well as data-flow logic. The results of these tests are reported and summarized in Chapter 6. One interesting result is that the algorithms seem to be balanced in efficiency; none dominates the others in the percentage of computation time used.

Finally, in Chapter 7 we briefly recapitulate the new ideas incorporated into ESPRESSO-II, and document their effectiveness through comparisons with existing logic minimizers. We conclude by suggesting directions for further research.

Chapter 2

BASIC DEFINITIONS

In this section we provide a framework of concise definitions for use in the entire sequel. The definitions will be informal, with the object being defined appearing in bold type.

Let $B = \{0,1\}$, $Y = \{0,1,2\}$. A **logic** (Boolean, switching) **function** $\mathcal{F}$ in n input variables, $x_1,..., x_n$, and m output variables, $y_1,..., y_m$, is a function

$$\mathcal{F}: B^n \rightarrow Y^m$$

where $x = [x_1,..., x_n] \in B^n$ is the **input** and $y = [y_1,..., y_m]$ is the **output** of $\mathcal{F}$. Note that in addition to the usual values of 0 and 1, the outputs y_i may also assume the **don't-care value** 2. We shall use only the symbol $\mathcal{F}$ to refer to such **incompletely specified** logic functions. A **completely specified function** f is a logic function taking values in $\{0,1\}^m$, i.e., all the values of the input map into 0 or 1 for all the components of f. For each component of $\mathcal{F}$, $\mathcal{F}_i$, $i = 1,..., m$, we can define: the **on-set**, $X_i^{ON} \subseteq B^n$, the set of input values x such that $\mathcal{F}_i(x) = 1$, the **off-set**,

X_i^{OFF}, the set of values such that $ff_i(x) = 0$ and the **don't-care** set, $X_i^{DC} \subseteq B^n$, the set of values such that $ff_i(x) = 2$. A logic function with m=1 is called a **single-output** function, while if m > 1, it is called a **multiple-output** function.

Many representations of a logic function are possible. The most straightforward one is the **tabular form or truth table.** In this form, the value of the outputs are specified for each possible combination of the inputs. For example, let $ff: B^3 \rightarrow Y^2$ be specified by

x_1	x_2	x_3	y_1	y_2
0	0	0	1	1
0	0	1	1	0
0	1	0	0	1
0	1	1	0	1
1	0	0	1	0
1	0	1	1	2
1	1	0	1	1
1	1	1	2	1

(2.1.1)

For this function, $X_1^{ON} = \{[0,0,0], [0,0,1], [1,0,0], [1,0,1], [1,1,0]\}$ $X_1^{OFF} = \{[0,1,0], [0,1,1]\}$, and $X_1^{DC} = \{[1,1,1]\}$. This tabular representation can be mapped into a geometrical representation by making use of Boolean n-cubes [ROT 80]. For our example, consider a cube in the Boolean 3-space as shown in Figure 2.1a Each vertex or point of this cube represents a value of the input. We can represent ff_1 and ff_2 by building two cubes in two Boolean 3-spaces as shown in Figure 2.1b and 2.1c. In general, a logic function with m output components, and n inputs has a geometric representation consisting of labeled n-cubes in m Boolean n-spaces. The **sets** X_i^{ON}, X_i^{OFF} and X_i^{DC} are in one-to-one correspondence with sets of **vertices** V_i^{ON}, V_i^{OFF} and V_i^{DC}, in the n-cubes of the Boolean spaces.

2.1 Operations on logic functions.

In the case of multiple output logic, the usual Boolean operations are performed componentwise on the outputs.

The **complement** of a completely specified logic function $f: B^n \rightarrow B^m$, $\bar{f}$, is a completely specified logic function, $\bar{f}: B^n \rightarrow B^m$, such that its components, $\bar{f}_1, \ldots, \bar{f}_m$, have their on-sets equal to the

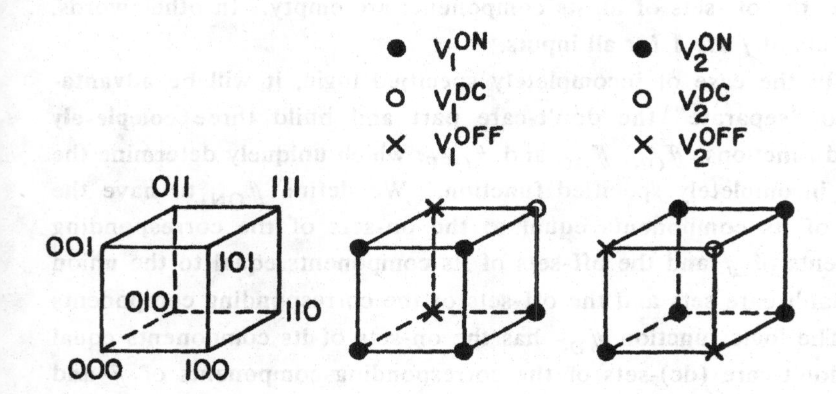

Figure 2.1 Covers of a 2-output Function

off-sets of the corresponding components of f. Their off-sets are equal to the on-sets of the corresponding components of f.

The **intersection** or **product** of two completely specified logic functions f and g, $h = f \bullet g$ or $f \cap g$, is defined to be the completely specified logic function h whose components, h_i, have on-sets equal to the intersection of the on-sets of the corresponding components of f and g.

The **difference** between two completely specified logic functions f and g, $h = f - g$, is a completely specified logic function h given by the intersection of f with the complement of g. The on-sets of the components of h are the elements of the on-sets of the corresponding components of f that are not in the on-set of the corresponding components of g.

The **union** or **sum** of two completely specified logic functions, f and g, f: $B^n \to \{0,1\}^m$, g: $B^n \to \{0,1\}^m$, is a completely specified logic function, $h = f \cup g$, $(h = f + g)$, such that the on-sets of the components of h, h_i, are the union of the on-sets of f_i and g_i.

A completely specified logic function f is a **tautology,** written $f \equiv 1$, if the off-sets of all its components are empty. In other words, the outputs of f are 1 for all inputs.

In the case of incompletely specified logic, it will be advantageous to "separate" the don't-care part and build three **completely** specified functions, ff_{ON}, ff_{DC} and ff_{OFF}, which uniquely determine the original incompletely specified function. We define ff_{ON} to have the on-sets of its components equal to the on-sets of the corresponding components of ff and the off-sets of its components equal to the union of the don't-care sets and the off-sets of the corresponding components of ff. The logic function ff_{DC} has the on-sets of its components equal to the don't-care (dc)-sets of the corresponding components of ff and the off-sets of its components equal to the union of the on-sets and the off-sets of the corresponding components. Finally, ff_{OFF} has the on-sets of its components equal to the off-sets of the corresponding components of ff and the off-sets of its components equal to the union of the on-sets and the dc-sets of the corresponding components of ff.

We will use the notation (f,d,r) to refer to the triple $(ff_{ON}, ff_{DC}, ff_{OFF})$ determined by ff. In our algorithms, we carry out our computations using these functions. Notice that $f \cup d \cup r$ covers the m Boolean spaces associated with ff. That is, for each component i, $1 \leq i \leq m$, the on-sets of f_i, d_i and r_i are mutually disjoint and their union is the entire vertex set of the Boolean n-cube. Thus, for each i, the on-sets of f_i, d_i and r_i partition the vertex set of the Boolean n-cube associated with component i. Hence $f \cup d \cup r$ is a tautology.

2.2 Algebraic representation of a logic function.

Consider ff_i, a component of ff. An algebraic representation of ff_i, f_i, is a Boolean expression [ROT 80] that evaluates to 1 for all inputs in X_i^{ON}, evaluates to 0 for all inputs in X_i^{OFF}, and evaluates *either* to 0 or 1 for all inputs in X_i^{DC}. (We "don't care" what values the function assumes on the don't-care set X_i^{DC}). An algebraic representation of $\mathcal{ff}$, f (sometimes denoted as f ($\mathcal{ff}$)), is a set of m Boolean expressions that are algebraic representations of the m components of $\mathcal{ff}$.

An algebraic representation of $\mathcal{ff}$ can be built by inspection from either the truth table or the geometric representation of $\mathcal{ff}$. For example, the algebraic representation of the ith component of $\mathcal{ff}$ can be built as follows. Consider every row of the truth table that has a 1 in the ith output position. Form a Boolean product (logical "and") of the n input variables, where x_j appears complemented if the corresponding value of the input variable in the row of the truth table is 0, uncomplemented if it is 1. This product obviously evaluates to 1 for the input combinations corresponding to the row of the truth table and to 0 for all other input combinations. If we form the Boolean sum (logical "or") of all the product terms so obtained, we obtain an algebraic representation for $\mathcal{ff}_i$. In our example (2.1.1), we have

$$\begin{aligned} f_1 &\equiv \bar{x}_1\bar{x}_2\bar{x}_3 + \bar{x}_1\bar{x}_2x_3 + x_1\bar{x}_2\bar{x}_3 + x_1\bar{x}_2x_3 + x_1x_2\bar{x}_3 \\ f_2 &\equiv \bar{x}_1\bar{x}_2\bar{x}_3 + \bar{x}_1x_2\bar{x}_3 + \bar{x}_1x_2x_3 + x_1x_2\bar{x}_3 + x_1x_2x_3. \end{aligned} \tag{2.2.1}$$

Note that these algebraic representations are in sum-of-products form. We shall focus on this form, since it can be used to represent any function, it is straightforward to implement with a PLA and it is easy to manipulate. Note also that many different algebraic representations of a logic function are possible. In particular, we have the freedom of assigning the don't-care inputs to evaluate to either 0 or to 1. For each decision we make, we obtain a different representation. For example, if we assign *d* of $\mathcal{ff}_2$ to 1, the algebraic representation of $\mathcal{ff}_2$ has one more product term, $x_1\bar{x}_2x_3$.

An algebraic representation of $\mathcal{ff}$ can be manipulated to yield more compact representations. For example, applying standard rules of Boolean algebra, we obtain

$$\begin{aligned} f_1 &= \bar{x}_2 + x_1\bar{x}_3 \\ f_2 &= x_2 + \bar{x}_1\bar{x}_3 \end{aligned} \tag{2.2.2}$$

Note that this representation is **different** from the one of (2.2.1). However, it is equivalent in the sense that the two components of (2.2.2) are 1 if and only if the components of (2.2.1) are 1. Also note that given a set of algebraic expressions, **f**, they uniquely determine a completely specified function $f(\mathbf{f})$. We will say that two algebraic representations **f**, **g**, are **equivalent** if they determine the same function f. For a single function f we may have many equivalent algebraic representations. We will also occasionally use the less cumbersome alphabetic notation, in which we replace $x_1, x_2, x_3, ...$ by A, B, C, ... In this notation, (2.2.2) becomes

$$\begin{aligned} f_1 &= \overline{B} + A\overline{C} \\ f_2 &= B + \overline{A}\overline{C} \end{aligned} \quad (2.2.3)$$

The operations of complementation, union, sum, intersection and product are defined for algebraic representations simply by applying the usual rules for Boolean manipulation. (Of course the outcome of such an operation is only well-defined up to the equivalence mentioned above.) For example, the **complement** of **f**, $\overline{\mathbf{f}}$, is obtained by applying the rules of complementation of Boolean expressions, and $\overline{\mathbf{f}}$ is an algebraic representation of $\overline{f}$.

Each of the product terms of a sum-of-products algebraic representation of f also determines a logic function. For example, referring to (2.2.2), $\overline{x}_2$ determines a logic function that is 1 on the vertices with coordinates (0,0,0), (1,0,1), (0,0,1), (1,0,0) of the Boolean 3-cube; $x_1\overline{x}_3$ determines a logic function that is 1 on the vertices with coordinates (1,0,0) and (1,1,0). Note that such vertex sets always form a cube, a 2-dimensional cube in the case of $\overline{x}_2$, and a 1-dimensional cube in the case of $x_1\overline{x}_3$. Thus a one-to-one correspondence between all possible product terms and k-dimensional cubes, $k \leq n$, in the Boolean n-space can be established.

2.3 Cubes and covers.

A **cube** in the Boolean n-space associated with a multiple-output function can be specified by its vertices and by an index indicating to which components of f it belongs. This information can be given in compact form. Let p be a product term associated with an algebraic sum of products expression of a logic function with n inputs and m outputs. Then a **cube** p is specified by a row vector $c = [c_1, \ldots, c_n, c_{n+1}, \ldots, c_{n+m}]$ where

$$c_i = \begin{cases} 0 & \text{if } x_i \text{ appears complemented in p,} \\ 1 & \text{if } x_i \text{ appears not complemented in p,} \quad (i = n+1, \ldots, n) \\ 2 & \text{if } x_i \text{ does not appear in p,} \\ 3 & \text{if p is not present in the algebraic representation of } f_{i-n} \quad (i = n+1, \ldots, n+m) \\ 4 & \text{if p is present in the algebraic representation of } f_{i-n} \end{cases} \tag{2.3.1}$$

For $p = \overline{x}_2$ in the above example, present only in the algebraic expression for f_1, we have

$$c = [2\ 0\ 2\ 4\ 3].$$

The **input part** of c, or **input cube** of c, I(c), is the subvector of c containing the first n entries of c. The **output part** of c, O(c), is the subvector of c containing the last m entries of c. A variable corresponding to a 2 in the input cube is referred to as an **input don't-care.**

Note that the input cube represents in a compact form the coordinates of the vertices of the cube corresponding to the product term. In the example, I(c) = [2 0 2] identifies all the vertices with a 0 in the second coordinate, i.e. the 2-cube (0,0,0), (1,0,0), (0,0,1), (1,0,1). The output cube, O(c) = [4 3] simply identifies the space to which the cube belongs. We have used the convention of 3 and 4 instead of 0 and 1 to avoid confusion between the input and output parts of the cube.

A set of cubes $\mathscr{C} = \{c^1, c^2, \ldots, c^k\}$ is said to be a **cover** for a logic function $\mathscr{f}$ with n inputs and m outputs, if, for j = 1,..., m, the set of input parts of the cubes that have a 4 in the jth position contain all the vertices corresponding to the on-set of $\mathscr{f}_j$ and none of the vertices of the off-set of $\mathscr{f}_j$, i.e. a cover represents the union of the onset and some arbitrary portion of the don't-care set.

Note that we have a one-to-one correspondence, defined by (2.3.1), between a set of cubes and a set of algebraic expressions. Because of the definitions of algebraic representation and cover, we have a one-to-one correspondence between covers and algebraic representations of a logic function as a sum of products. Therefore, we shall use the terms cover and algebraic representation of $\mathscr{f}$ interchangeably.

A set of cubes $\mathscr{C}$ determines a completely specified logic function, denoted by b($\mathscr{C}$).

In all our algorithms, we use a matrix representation of a cover (or of an algebraic representation). The matrix M($\mathscr{C}$) associated with a cover $\mathscr{C} = \{c^1, \ldots, c^k\}$ is the matrix obtained by stacking the row vectors representing each of the cubes of $\mathscr{C}$. For example, the matrix representation of $\mathscr{C}$ corresponding to (2.2.2) is:

$$F = M(\mathscr{C}) = \begin{matrix} 2 & 0 & 2 & 4 & 3 \\ 1 & 2 & 0 & 4 & 3 \\ 2 & 1 & 2 & 3 & 4 \\ 0 & 2 & 0 & 3 & 4 \end{matrix} \qquad (2.3.2)$$

We will call M($\mathscr{C}$) a **matrix representation** of a logic function f (matrices M($\mathscr{C}$) will be denoted M(f) to indicate explicitly this relation). Of course, any set of cubes, whether it is a cover or not, has a matrix representation.

As in the notation f(f), we shall use $f(\mathscr{C})$ or $f(M(\mathscr{C}))$ to represent the completely specified logic function induced by the cover $\mathscr{C}$ or the matrix M($\mathscr{C}$). Note that the distinctive feature of the symbols for such induced functions is lower case italics. Thus we sometimes use h(M) or $h(\mathscr{C})$.

We further define G = I (M($\mathscr{C}$)) and H = O(M($\mathscr{C}$)) to be the matrices corresponding to the input and output parts respectively of a cover $\mathscr{C}$. If $\mathscr{C}$ is the cover of a single output function we let H be the (empty) matrix of zero columns, since the output part would be redundant in that case.

In the sequel, it would be laborious to continually make the distinction between a cover and its matrix representation, so hence forth, we shall use these terms interchangeably when the context is clear.

Let $c = \{c_1,..., c_{n+m}\}$ and $d = \{d_1,..., d_{n+m}\}$ be two cubes. We say that c_j **contains** ($\supseteq$), does not contain ($\not\supseteq$), or **strictly contains** ($\supset$) d_j according to the following table:

c_j \ d_j	0	1	2	
0	$\supseteq$	$\not\supseteq$	$\not\supseteq$	
1	$\not\supseteq$	$\supseteq$	$\not\supseteq$	$1 \leq j \leq n$
2	$\supset$	$\supset$	$\supseteq$	

c_j \ d_j	3	4	
3	$\supseteq$	$\not\supseteq$	$n < j \leq n+m$
4	$\supset$	$\supseteq$	

Then cube c **contains** (covers) d, $c \supseteq d$, if each entry of c contains the corresponding entry of d; c **strictly contains** d, $c \supset d$, if c contains d and for at least one j, $c_j \supset d_j$. In geometrical terms, the cube represented by the input part of c must contain all the vertices of the cube represented by d and must be present in all Boolean spaces where c is present.

A **minterm** e^i is a cube whose input part does not contain any 2's and whose output part contains (m - 1) 3's and one 4 in position i. In geometrical terms, the input cube of a minterm is a vertex and the output part specifies that this vertex is present only in the ith Boolean n-space. Note that a minterm does not contain any cube other than itself and therefore we can think of a minterm as an atomic constituent of a cube. If a cube c contains minterm, e^i, $c \supseteq e^i$, we say that e^i is an **element** of c, $e^i \in c$. For example [1 1 1 4 3] is a minterm and is an element of [2 2 1 4 4].

Each cube can be decomposed into the set of all minterms that are elements of the cube. In our example, c = [2 2 1 4 4] can be decomposed into

$m^1 = [0\ 0\ 1\ 4\ 3]$, $m^2 = [0\ 0\ 1\ 3\ 4]$, $m^3 = [1\ 0\ 1\ 4\ 3]$,
$m^4 = [1\ 0\ 1\ 3\ 4]$, $m^5 = [0\ 1\ 1\ 4\ 3]$, $m^6 = [0\ 1\ 1\ 3\ 4]$,
$m^7 = [1\ 1\ 1\ 4\ 3]$, $m^8 = [1\ 1\ 1\ 3\ 4]$.

A set of cubes $\mathscr{C} = \{c^1,..., c^k\}$ **covers** a cube c (we write $c \subseteq \mathscr{C}$) if each of the minterms of c is covered by at least one cube of $\mathscr{C}$.

There are three other cubes of interest in our algorithms: the cube u^j representing the **universe** in the jth Boolean space, the cube U representing the total universe, and the cube x^j representing the positive half-space of the literal x_j:

	1 2 ... j ... n	n+1	n+2	n+j	n+m
u^j =	2 2 ... 2 ... 2	3	3	... 4 ...	3
U =	2 2 ... 2 ... 2	4	4	... 4 ...	4
x^j =	2 2 ... 1 ... 2	4	4	... 4 ...	4
$\bar{x}^j$ =	2 2 ... 0 ... 2	4	4	... 4 ...	4

The **intersection** or **and** or **product** of two cubes c and d, written $c \cap d$, or cd, is a cube e. The entries e_i of the cube are obtained from the entries of c and d according to the following table

c_i \ d_i: $\cap$	0	1	2
0	0	ϕ	0
1	ϕ	1	1
2	0	1	2

$1 \le i \le n$

c_i \ d_i: $\cap$	3	4
3	3	3
4	3	4

$n < i \le n+m$

When there is an index i such that c_i **and** d_i give ϕ, the cube e is said to be the **empty cube.** If the output part of e has all 3's, e is also the empty cube. Geometrically, the cube obtained by intersecting two cubes, is a cube whose input part corresponds to the vertices that are common to c and d, and whose output part specifies that this cube is present in the Boolean n-spaces in which both c and d are present. Thus, if the two input cubes have no vertices in common or if they are not both present in **any** of the m Boolean n-spaces, the intersection is empty, the two cubes are said to be **orthogonal cubes,** and we write $c \cap d = \phi$. (The empty cube is not really a cube by our definition, but we introduce it as a matter of convenience.)

The **intersection of two sets of cubes** is the set of cubes obtained by performing the pairwise intersection of all the cubes in the two sets. The sets of cubes are **orthogonal** if their intersection consists only of empty cubes. Note that any cover of f has to be orthogonal to any cover of $\bar{f}$.

The **union** or **or** or **sum** of two cubes c and d written $c \cup d$, $c + d$, is the set of vertices covered by the input part of either c or d in the Boolean n-space where they are present. If we use a matrix representation of the cubes, $c \cup d$ is the matrix formed by the two rows corresponding to c and d respectively.

The **distance between two cubes** c and d, $\delta(c, d)$, is equal to the number of input "conflicts", i.e. of input cube entries that have intersection equal to ϕ, plus 1 if the intersection of the output parts gives an output part of all 3's. That is,

$$\delta(c, d) = \delta(I(c), I(d)) + \delta(O(c), O(d))$$

where

$$\delta(I(c), I(d)) = |\{j \mid c_j \cap d_j = \phi\}|$$

and

$$\delta(O(c), O(d)) = \begin{cases} 0 & \text{if } c_j \cap d_j = 4, \text{ some } j > n, \\ 1 & \text{otherwise} \end{cases}$$

The **consensus** of **two cubes** c and d (written $e = c \odot d$), is a cube defined as follows. If $\delta(c, d) \neq 1$ then

$$e = \begin{cases} c \cap d & \text{if } \delta(c, d) = 0, \\ \phi & \text{if } \delta(c, d) \geq 2. \end{cases}$$

If $\delta(I(c), I(d)) = 1$ and $\delta(O(c), O(d)) = 0$, then the entries of e are given by

$$e_l = \begin{cases} c_l \cap d_l & \text{if } c_l \cap d_l \neq \phi, \\ 2 & \text{otherwise,} \end{cases}$$

that is, the conflicting input-part is raised to 2 and the intersection is taken. If $\delta(I(c), I(d)) = 0$ and $\delta(O(c), O(d)) = 1$ then

$$e_l = \begin{cases} c_l \cap d_l & \text{for } 1 \le l \le n, \\ 4 & \text{if } c_l \text{ or } d_l = 4, \text{ for } n < l \le n+m, \\ 3 & \text{otherwise.} \end{cases}$$

Proposition 2.3.1. The consensus of two cubes a and b, $p = a \odot b$, is contained in $a \cup b$. Furthermore, if $a \odot b \neq \phi$, it contains minterms of both a and b.

Proof. If $\delta(a,b) = 0$, then $p = a \cap b$ and the Proposition holds. If $\delta(I(a), I(b)) = 1$, then let k be the conflicting input part. Then

$$I(p) = \{a_1 \cap b_1, \dots, a_k \cup b_k, \dots, a_n \cap b_n\}, \; O(p) = O(a) \cap O(b).$$

All minterms of p are either in p' defined by

$$I(p') = \{a_1 \cap b_1, \dots, a_k, \dots, a_n \cap b_n\}; \; O(p') = O(a) \cap O(b)$$

or in p'' defined by

$$I(p'') = \{a_1 \cap b_1, \dots, b_k, \dots, a_n \cap b_n\}; \; O(p'') = O(a) \cap O(b).$$

But $a \supset p'$, $b \supset p''$ and the Proposition holds.

If $\delta(O(a), O(b)) = 1$, then all minterms of p are either in p' defined by

$$I(p') = I(a) \cap I(b); \; O(p') = O(a)$$

or in p'' defined by

$$I(p'') = I(a) \cap I(b); \; O(p'') = O(b).$$

Then $a \supset p'$, $b \supset p''$ and the Proposition holds. Finally, if $\delta(a, b) > 1$ then $p = \phi \subset a$ and also $p \subseteq b$. ■

This proposition states that $c = a \odot b$ is a cube with one "leg" in a and one in b; c is a sort of bridge between a and b.

The geometric interpretation of consensus is illustrated in Figure 2.3.1. Cubes $c_1 - c_6$ have "encircled" vertices in Figure 2.3.1, which illustrates a Boolean function with 2 outputs and 3 inputs. The cubes

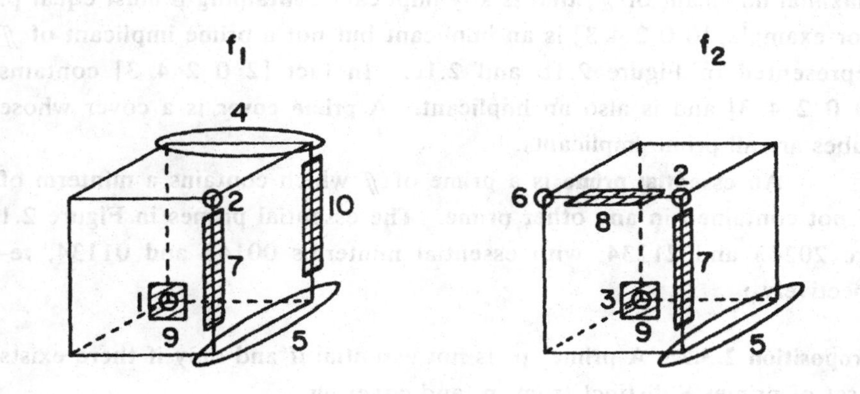

Figure 2.3.1. Consensus Between 2-Output Cubes in 3-Space.

$c_7 = c_2 \odot c_5$, $c_8 = c_2 \odot c_6$, $c_9 = c_1 \odot c_3$ and $c_{10} = c_5 \odot c_4$ are indicated by the shaded rectangles. Note $c_2 \odot c_3 = c_1 \odot c_2 = c_3 \odot c_6 = c_5 \odot c_6 = \phi$.

A set of cubes directly represents a completely specified logic function. A set of cubes may **cover** an incompletely specified logic function.

The **complement** of a set of cubes $\mathscr{C}$, $\overline{\mathscr{C}}$, is a set of cubes which covers the complement of the logic function corresponding to $\mathscr{C}$. The **difference** between two sets of cubes, $\mathscr{C} - \mathscr{H}$, is a set of cubes that covers $\mathscr{C} \cap \overline{\mathscr{H}}$.

A cube p is said to be an **implicant** of $\mathscr{F} = (f,d,r)$ if it has empty intersection with the cubes of a representation of r. In our example, c^2 of F in (2.3.2), [1 2 0 4 3], is an implicant of $\mathscr{F}$, while [0 2 1 3 4] is not, since it contains the vertex (0 1 1) in the Boolean

3-space representing $\mathscr{f}_2$ that is in the off-set of $\mathscr{f}_2$ as shown in Figure 2.1c.

A **prime implicant** or **prime cube** or simply **prime** of $\mathscr{f}$, $\tilde{p}$, is a maximal implicant of $\mathscr{f}$; that is any implicant containing $\tilde{p}$ must equal $\tilde{p}$. For example, [0 0 2 4 3] is an implicant but not a prime implicant of $\mathscr{f}$ represented in Figure 2.1b and 2.1c. In fact [2 0 2 4 3] contains [0 0 2 4 3] and is also an implicant. A **prime cover** is a cover whose cubes are all prime implicants.

An **essential prime** is a prime of $\mathscr{f}$ which contains a minterm of $\mathscr{f}$ not contained in any other prime. The essential primes in Figure 2.1 are 20243 and 21234, with essential minterms 00143 and 01134, respectively.

Proposition 2.3.2: A prime p is not essential if and only if there exists a set of primes S distinct from p and covering p.

A cover $\mathscr{C}$ such that no proper subset of $\mathscr{C}$ is also a cover is said to be **irredundant** or **minimal.**

The above definitions can be stated for a set of cubes. A set of cubes, $\mathscr{C}$ is said to be **irredundant** or **minimal** if no cube in $\mathscr{C}$ is covered by the set of other cubes of $\mathscr{C}$. A cube $c \in \mathscr{C}$ is said to be a prime cube, if no cube properly containing c is entirely covered by $\mathscr{C}$.

A weaker notion of minimality is that of **minimality with respect to single cube containment.** A set of cubes $\mathscr{C}$ has this property if $a \not\subseteq b$ for any distinct cubes $a, b \in \mathscr{C}$.

Chapter 3

DECOMPOSITION AND UNATE FUNCTIONS

Except for the procedure EXPAND, all our algorithms are based on a single fundamental strategy: a recursive divide and conquer. Examples of the use of this strategy applied to logic function manipulation can be found in papers by Morreale [MOR 70] and by Hong and Ostapko [HON 72]. The decomposition is based on the Shannon expansion [SHA 48]. The Shannon expansion makes use of the **cofactor** of a logic function. Since we are dealing primarily with matrix representations and covers, we will give the definition of cofactor for representations.

3.1 Cofactors and Shannon expansion

Given a set of cubes $\mathcal{G} = \{c^1, \ldots, c^\ell\}$ and a cube p, all with n entries in the input part and m entries in the output part, the **cofactor** of $\mathcal{G}$ with respect to p, $\mathcal{G}_p$ is a set of cubes (possibly empty) obtained by computing the cofactor of each of the cubes in the cover $\mathcal{G}$. We define the cofactor of c^i with respect to p as the cube with components

$$(c_p^i)_k = \begin{cases} \phi & \text{if } c^i \cap p = \phi \\ 2 & \text{if } p_k = 0 \text{ or } 1 \\ 4 & \text{if } p_k = 3 \\ c_k^i & \text{otherwise.} \end{cases} \tag{3.1.0}$$

For example, the cofactor of

$$G = M(\mathscr{G}) = \begin{bmatrix} 1 & 1 & 0 & 2 & 4 & 4 \\ 0 & 1 & 2 & 0 & 4 & 4 \\ 1 & 1 & 1 & 1 & 4 & 3 \end{bmatrix} \tag{3.1.1}$$

with respect to $p = [1\ 1\ 2\ 2\ 4\ 3]$ is given by

$$G_p = \begin{bmatrix} 2 & 2 & 0 & 2 & 4 & 4 \\ 2 & 2 & 1 & 1 & 4 & 4 \end{bmatrix}.$$

In our algorithms, we will use cofactors with respect to the cube x^j. For example, the cofactor of G of (3.1.1) with respect to the cube $x^4 = [2\ 2\ 2\ 1\ 4\ 4]$ is

$$G_{x^4} = \begin{bmatrix} 1 & 1 & 0 & 2 & 4 & 4 \\ 1 & 1 & 1 & 2 & 4 & 3 \end{bmatrix}.$$

If we take the cofactor of G with respect to $\bar{x}^4 = [2\ 2\ 2\ 0\ 4\ 4]$, we obtain

$$G_{\bar{x}^4} = \begin{bmatrix} 1 & 1 & 0 & 2 & 4 & 4 \\ 0 & 1 & 2 & 2 & 4 & 4 \end{bmatrix}.$$

The cofactor, g_p, of a completely specified logic function g with respect to a cube p is defined to be the logic function corresponding to the cofactor of a matrix representation G of g with respect to p, i.e., $g_p = b(G_p)$.

Let us now consider **g**, an algebraic representation of the logic function corresponding to G, and $\mathbf{g}_{x_j}$ and $\mathbf{g}_{\bar{x}_j}$, algebraic representations of the logic functions corresponding to G_{x^j} and to $G_{\bar{x}^j}$. Then the **Shannon expansion** of **g** is

$$\mathbf{g} = x_j\, \mathbf{g}_{x_j} + \bar{x}_j\, \mathbf{g}_{\bar{x}_j}, \tag{3.1.2}$$

and $\mathbf{g}_{x_j}$, is called the cofactor of **g** with respect to the variable x_j. If we form the following cover from G_{x^j} and $G_{\bar{x}^j}$

$$x^j\, G_{x^j} + \bar{x}^j\, G_{\bar{x}^j}$$

we obtain a set of cubes that give another representation of the function **g** corresponding to G. In our example

$$x^4 G_{x^4} = \begin{bmatrix} 1 & 1 & 0 & 1 & 4 & 4 \\ 1 & 1 & 1 & 1 & 4 & 3 \end{bmatrix} \; ; \quad \bar{x}^4 G_{\bar{x}^4} = \begin{bmatrix} 1 & 1 & 0 & 0 & 4 & 4 \\ 0 & 1 & 2 & 0 & 4 & 4 \end{bmatrix} \; ;$$

and

$$\hat{G} \equiv x^4 G_{x^4} + \bar{x}^4 G_{\bar{x}^4} = \begin{matrix} 1 & 1 & 0 & 1 & 4 & 4 \\ 1 & 1 & 0 & 0 & 4 & 4 \\ 1 & 1 & 1 & 1 & 4 & 3 \\ 0 & 1 & 2 & 0 & 4 & 4 \end{matrix}$$

$\hat{G}$ covers the same vertices as G and hence corresponds to the same logic function g, i.e. G and $\hat{G}$ are logically equivalent. In terms of the algebraic representation of g, **g**, the Shannon expansion corresponds to "collecting" common terms from the product terms of the algebraic representation. In our example

$$\mathbf{g} = \begin{bmatrix} \mathbf{g}_1 \\ \mathbf{g}_2 \end{bmatrix} = \begin{bmatrix} \bar{x}_1 x_2 \bar{x}_4 + x_1 x_2 x_3 x_4 + x_1 x_2 \bar{x}_3 \\ x_1 x_2 \bar{x}_3 + \bar{x}_1 x_2 \bar{x}_4 \end{bmatrix}$$

The Shannon expansion yields

$$\begin{aligned} \mathbf{g} &= x_4 \mathbf{g}_{x_4} + \bar{x}_4 \mathbf{g}_{\bar{x}_4} \\ &= x^4 \begin{bmatrix} x_1 x_2 x_3 + x_1 x_2 \bar{x}_3 \\ x_1 x_2 \bar{x}_3 \end{bmatrix} + \bar{x}_4 \begin{bmatrix} \bar{x}_1 x_2 + x_1 x_2 \bar{x}_3 \\ x_1 x_2 x_3 + \bar{x}_1 x_2 \end{bmatrix}. \end{aligned} \tag{3.1.3}$$

Note that the equality in (3.1.3) is intended to mean that $x_4 \mathbf{g}_{x_4} + \bar{x}_4 \mathbf{g}_{\bar{x}_4}$ has the same truth table as **g** i.e., is **logically equivalent to g.**

Recursive algorithms based on the Shannon expansion rely on the following fundamental propositions.

Proposition 3.1.1a: Let **f** and **g** be algebraic representations of completely specified Boolean functions. Then the following operations commute:

1. Intersection of two functions and cofactor operations, i.e.

$$(fg)_{x_j} = (f_{x_j} g_{x_j}) \tag{3.1.4a}$$

2. Complementation of a function and cofactor operations, i.e.

$$(\bar{f})_{x_j} = \overline{(f_{x_j})} \tag{3.1.4b}$$

Proof: We make use of the Shannon expansion to prove these results. For the intersection, since $x^j \cup \bar{x}^j = U$, $f = x_j f_{x_j} + \bar{x}_j f_{\bar{x}_j}$ and $g = x_j g_{x_j} + \bar{x}_j g_{\bar{x}_j}$, we have

$$fg = (x_j f_{x_j} + \bar{x}_j f_{\bar{x}_j})(x_j g_{x_j} + \bar{x}_j g_{\bar{x}_j}) = x_j(f_{x_j} g_{x_j}) + \bar{x}_j(f_{\bar{x}_j} g_{\bar{x}_j}) \tag{3.1.5}$$

But the function fg can also be expanded as

$$fg = x_j(fg)_{x_j} + \bar{x}_j(fg)_{\bar{x}_j} \tag{3.1.6}$$

Since $(fg)_{x_j}$ and f_{x_j}, g_{x_j} are both independent of x_j, equating (3.1.5) and (3.1.6) yields $(fg)_{x_j} = f_{x_j} g_{x_j}$.

For complementation,

$$\begin{gathered}\bar{f} = \overline{(x_j f_{x_j} + \bar{x}_j f_{\bar{x}_j})} = (\bar{x}_j + \overline{(f_{x_j})})(x_j + \overline{(f_{\bar{x}_j})}) = \\ \bar{x}_j\overline{(f_{\bar{x}_j})} + x_j\overline{(f_{x_j})} + \overline{(f_{x_j})}\ \overline{(f_{\bar{x}_j})} = \bar{x}_j\overline{(f_{\bar{x}_j})} + x_j\overline{(f_{x_j})}\end{gathered} \tag{3.1.7}$$

The last equality occurs because the vertices covered by $\overline{(f_{x_j})}\ \overline{(f_{\bar{x}_j})}$ are all covered by $\bar{x}_j\overline{(f_{\bar{x}_j})} + x_j\overline{(f_{x_j})}$. Since $\bar{f}$ has the Shannon expansion with respect to x_j,

$$\bar{f} = x_j(\bar{f})_{x_j} + \bar{x}_j(\bar{f})_{\bar{x}_j}. \tag{3.1.8}$$

Since $(\bar{f})_{x_j}$ and $\overline{(f_{x_j})}$ are both independent of x_j, equating (3.1.7) and (3.1.8) yields $(\bar{f})_{x_j} = \overline{(f_{x_j})}$. ■

Proposition 3.1.1b: Let $\mathscr{G} = \{c^i\}$ be a cover of cubes and let p be a cube. Then

$$p \cap \mathscr{G} \equiv p \cap \mathscr{G}_p.$$

Proof: By the definition of (3.1.0) we have, assuming $p \cap c^i \neq \phi$,

$$(p \cap c_p^i)_k = \begin{cases} p_k, & \text{if } p_k = 0 \text{ or } p_k = 1 \text{ or } p_k = 3 \\ (c^i)_k, & \text{if } p_k = 2 \text{ or } p_k = 4. \end{cases}$$

This is because $(c_p^i)_k = 2$ if $p_k = 0$ or 1. But by the definition table of intersection given in Section 2.3, again assuming $p \cap c^i \neq \phi$,

$$(p \cap c^i)_k = \begin{cases} p_k, & \text{if } p_k = 0 \text{ or } p_k = 1 \text{ or } p_k = 3 \\ (c^i)_k, & \text{if } p_k = 2 \text{ or } p_k = 4. \end{cases}$$

Thus either $p \cap \mathscr{G} = \phi$ and $\mathscr{G}_p = \phi$ or, for each cube $p \cap c^i$ in $p \cap G$, we have

$$(p \cap c^i)_k \equiv (p \cap c_p^i)_k, \text{ for all k.}$$ ■

These results lead to the following simple but useful observation, relating the tautology test with testing if a cube c is is covered by a set of cubes $\mathscr{C}$.

Theorem 3.1.2. A set of cubes $\mathscr{C}$ covers a cube c if and only if $\mathscr{C}_c$, the cofactor of $\mathscr{C}$ with respect to c, is a tautology.

Proof: Clearly $\mathscr{C}$ covers c if and only if

$$\mathscr{C} \cap c = c. \tag{3.1.9}$$

Suppose (3.1.9) is valid. Then taking the cofactor of both sides implies $(\mathscr{C} \cap c)_c = c_c$. Using the commutivity of intersection and cofactor, (cf., Proposition 3.1.1a) and the fact that $c_c \equiv 1$, we have $\mathscr{C}_c \equiv 1$. Conversely, suppose $\mathscr{C}_c \equiv 1$, then $\mathscr{C}_c \cap c = c$. By Proposition 3.1.1b, $\mathscr{C}_c \cap c = \mathscr{C} \cap c$ proving the result. ■

Remark. In the sequel we will not distinguish between the cofactor, f_p, of a Boolean function f with respect to a cube p, and the cofactor, $\mathbf{f}_p$, of an algebraic representation **f** (f) with respect to p.

Remark. From now on we will denote $\overline{(f_{x_j})}$ simply as $\bar{f}_{x_j}$. Note that (3.1.4b) and (3.1.8) suggest a recursive divide and conquer procedure to compute the complement of a function and the intersection of two functions. This recursive strategy can also be applied to another operation that will be used over and over in our algorithms: checking if a given logic function is a **tautology.**

Proposition 3.1.2. Let $f = x_j f_{x_j} + \bar{x}_j f_{\bar{x}_j}$ be the Shannon expansion of a completely specified logic function f. Then $f \equiv 1$ (i.e. f is a tautology) if and only if $f_{x_j} \equiv 1$ and $f_{\bar{x}_j} \equiv 1$.

Proof: If part. If f_{x_j} and $f_{\bar{x}_j}$ are tautologies, the values of their outputs are 1 for all values of the inputs. Hence $x_j f_{x_j} + \bar{x}_j f_{\bar{x}_j} = x_j + \bar{x}_j \equiv 1$.

Only if part. Suppose one of the cofactors of f, say f_{x_j}, is not a tautology. Then there exists an input combination so that one of the outputs of f_{x_j}, say y_i, is 0. Since f_{x_j} is independent of x_j, an input combination with $x_j = 1$ can be selected so that the output of y_i of $x_j f_{x_j}$ is 0. For this input combination, the output of $f = x_j f_{x_j} + \bar{x}_j f_{\bar{x}_j}$ corresponding to y_i must also be 0 and therefore f is not a tautology. ■

These results provide the basis for the general paradigm that applies to all of the above operations, intersection, complementation and tautology. This paradigm can be stated simply in two phases:

1. apply the operation to the two cofactors;

2. merge the results.

For example, given an operation involving two functions f and g, the paradigm is

$$\text{operate } (f,g) = \text{merge } (x_j \text{ operate } (f_{x_j}, g_{x_j}), \ \bar{x}_j \text{ operate } (f_{\bar{x}_j}, g_{\bar{x}_j}))$$

First we focus on the **merging process.**

3.2 Merging

The recursion process creates a binary tree and the results at each node will be a merging of the results obtained from the two subtrees below it. For tautology checking the merging is obvious by Proposition 3.1.2. Not so for complementation or for the reduction operation that we will describe in Section 4.6. In both these cases the two results obtained by operating on the cofactors are in the form of a set of cubes or a cover of a logic function. Let h_0 and h_1 be the logic functions corresponding to the results obtained by the subtrees below a node of the tree. The merging gives us the "resulting" function as $h = \bar{x}_j h_0 + x_j h_1$. Since we work with covers and their matrix representation, the result computed by our algorithms will be, if the Shannon expansion is applied verbatim, $\mathcal{H} = \bar{x}^j \mathcal{H}_0 + x^j \mathcal{H}_1$, where $\mathcal{H}$, $\mathcal{H}_0$, $\mathcal{H}_1$ are covers of h, h_0, h_1 respectively.

Our goal is to keep the result in as compact a form as possible. Let us assume for example, that $\mathcal{H}_0$ and $\mathcal{H}_1$ are prime irredundant covers of h_0 and h_1. Our objective is then to preserve this property to the next stage; that is, to obtain a cover $\tilde{H}$ of h which is itself prime and irredundant.

To accomplish this objective the first task is to make all the cubes in $\mathcal{H}$ prime. To do that, we pick one cube of $\mathcal{H}$ at a time and make it as "large" as possible, compatible with the requirements that it must not contain any vertex that is not in $\mathcal{H}$. Then, to make the cover irredundant, we have to remove from $\mathcal{H}$ all the redundant cubes, i.e. the cubes whose vertices are covered by the remainder of the cover.

Because of the assumption that $\mathcal{H}_0$ and $\mathcal{H}_1$ are prime irredundant covers, and of the form of the Shannon expansion, making the cubes of $\mathcal{H}$ as large as possible, is conceptually simple. In order to make a cube larger, one has to either change a 0 or a 1 in the input cube, into a 2, or change a 3 in the output part into a 4. The cube obtained by this operation will strictly contain the original cube. Since $\mathcal{H}_0$ and $\mathcal{H}_1$ are prime covers there is no hope of changing a 1 or a 0, in a position different than j, into a 2, or of changing a 3 into a 4 in the output part. The only possibility of enlarging cubes in $\mathcal{H}$ is to raise the values in the jth coordinate of the cubes in $\bar{x}^j \mathcal{H}_0$ and $x^j \mathcal{H}_1$, to a 2. For example, let us pick up a cube of $x^j \mathcal{H}_1$, say

$c = [c_1, ..., c_j = 1, ..., c_{n+m}]$. For this cube to become $\tilde{c} = [c_1, ..., 2, ..., c_{n+m}]$, $\hat{c} = [c_1, ..., 0, ..., c_{n+m}]$ must be covered by $\bar{x}^j \mathcal{H}_0$. Therefore, the checking is restricted only to $\bar{x}^j \mathcal{H}_0$ for the cubes in $x^j \mathcal{H}_1$, and restricted to $x^j \mathcal{H}_1$ for the cubes in $\bar{x}^j \mathcal{H}_0$.

Unfortunately, the operation of "raising" the cubes as well as the irredundant cover operation are rather expensive. Hence a faster approximate procedure is desired at the expense of obtaining only a "good" cover rather than a guaranteed minimal prime cover. In Figure 3.2.1, we give a procedure which implements two merging strategies: MERGE_WITH_IDENTITY and MERGE_WITH_CONTAINMENT. The first is faster and is invoked by a global switch CONTAIN, while the second is more powerful but takes more time. We will have occasion to use both strategies.

The procedure MERGE_WITH_IDENTITY checks only for identity between cubes in $\mathcal{H}_0$ and $\mathcal{H}_1$ while MERGE_WITH_CONTAINMENT checks additionally for containment by **a single** cube of $\mathcal{H}_0$ or $\mathcal{H}_1$. Therefore using these merging processes we cannot guarantee that $\mathcal{H}$ is a set of primes. Moreover, the only possible redundancy that is detected is when two cubes in $\mathcal{H}_0$ and $\mathcal{H}_1$ are found to be identical. Note that identity is always checked first as an initial filter, since it obviously reduces the cardinality of $\mathcal{H}_0$ and $\mathcal{H}_1$, which must be scanned for pairwise containment if CONTAIN=1. Identity can be checked efficiently using encoding and sorting in roughly O(n log n) time where $n = \{ |\mathcal{H}_0| + |\mathcal{H}_1| \}$. Straightforward pairwise containment is $O(|\mathcal{H}_0| \, |\mathcal{H}_1|)$; this procedure is heuristically improved by ordering the cubes in $\mathcal{H}_0$ and $\mathcal{H}_1$ according to the "size" of a cube, using the fact that a smaller cube cannot contain a larger one.

3.3 Unate Functions

For the recursive step of the paradigm, a variable x_j called the **splitting variable,** must be selected, so as to minimize the complexity of the solution of the subproblems involving the cofactors of the logic function with respect to $\bar{x}_j$ and x_j. It turns out that there are very efficient methods for tautology and reduction, as well as a good heuristic for complementation, if we restrict our attention to a particular class of logic functions: unate functions. Therefore, we select the splitting

```
Procedure MERGE_WITH_CONTAINMENT (𝒳_0, 𝒳_1)
/* Given two subcovers 𝒳_0 and 𝒳_1 obtained by the Shannon expansion
/* with respect to x_j, computes 𝒳, a cover obtained by merging 𝒳_0 and 𝒳_1.
/* The cubes of 𝒳_0 are denoted by 𝒳_0^i and the cubes of 𝒳_1, are denoted by 𝒳_1^ℓ
/* If the global switch CONTAIN is 0, then this procedure becomes
/* MERGE_WITH_IDENTITY.

    Begin
    k ← |𝒳_0|; p ← |𝒳_1|; 𝒳_2 ← φ
    for (i = 1,..., k)
        Begin
        for (ℓ = 1,..., p)
            Begin
            if (𝒳_0^i = 𝒳_1^ℓ)                /* If there are two identical cubes
                Begin                          /* in 𝒳_0 and 𝒳_1 remove them
                𝒳_2 ← 𝒳_2 ∪ {𝒳_0^i}            /* and put them in 𝒳_2.
                𝒳_0 ← 𝒳_0 − {𝒳_0^i}
                𝒳_1 ← 𝒳_1 − {𝒳_1^ℓ}
                End
            End
        End
    if (CONTAIN = 0) return (𝒳 ← x̄^j 𝒳_0 + x^j 𝒳_1 + 𝒳_2)
    k ← |𝒳_0|; p ← |𝒳_1|                      /* If the switch CONTAIN is
    for (i = 1,..., k)                         /* off, then return the current result.
        Begin                                  /* This defines the procedure
        for (ℓ = 1,..., p)                     /* MERGE_WITH_IDENTITY
            Begin
            if (𝒳_0^i ⊃ 𝒳_1^ℓ) then            /* If 𝒳_0^i contains 𝒳_1^ℓ, then remove
                Begin                          /* 𝒳_1^ℓ from 𝒳_1, and add it to 𝒳_2.
                𝒳_2 ← 𝒳_2 ∪ {𝒳_1^ℓ}
                𝒳_1 ← 𝒳_1 − {𝒳_1^ℓ}
                End
            else
                Begin                          /* If 𝒳_1^ℓ contains 𝒳_0^i then
                if (𝒳_1^ℓ ⊃ 𝒳_0^i)             /* remove 𝒳_0^i from 𝒳_0 and add it
                    Begin                      /* to 𝒳_2.
                    𝒳_2 ← 𝒳_2 ∪ {𝒳_0^i}
                    𝒳_0 ← 𝒳_0 − {𝒳_0^i}
                    End
                End
            End
        End
    return (𝒳 ← x̄^j 𝒳_0 + x^j 𝒳_1 + 𝒳_2)
    End
```

Figure 3.2.1

variable to make each of the cofactors as close as possible to a unate function. This choice is discussed in Section 3.4. In this section we develop the relevant properties of unate functions.

A logic function f is **monotone increasing (monotone decreasing)** in a variable x_j if changing x_j from 0 to 1 causes **all** the outputs of f that change, to increase also from 0 to 1 (from 1 to 0). A function that is either monotone increasing or monotone decreasing in x_j is said to be **monotone** or **unate** in x_j. A function is **unate (monotone)** if it is unate in all its variables. For example, the function $f = x_1\bar{x}_2 + \bar{x}_2x_3$ is unate since it is increasing in x_1 and x_3, and decreasing in x_2.

A cover $\mathscr{C}$ is monotone increasing (decreasing) in the variable x_j if all the cubes of $\mathscr{C}$ have either a 1 (a 0) or a 2 in position j. A cover is said to be unate if it is monotone in all the input variables.

Proposition 3.3.1: If a cover $\mathscr{C}(f)$ is unate in x_j, then f is unate in x_j.

Proof: A cover that is unate, say monotonically increasing in x_j, has an algebraic representation as a set of sum-of-products Boolean expression where x_j either does not appear (2 in the jth position of the corresponding cube) or appears uncomplemented (1 in the jth position of the corresponding cube). Then changing x_j from 0 to 1 can only "activate" the product terms where it appears and the proposition is proven. ■

Note that a function that is unate may have a cover that is nonunate. For example, the logic function induced by the matrix

$$F = \begin{bmatrix} 1 & 1 & 0 & 4 \\ 2 & 0 & 2 & 4 \end{bmatrix}$$

is unate. In fact, it is obviously unate in x_1 and x_3 by Proposition 3.3.1, but it is also unate in x_2. However, there is a cube with a 0 and a cube with a 1 in the second position, so the cover is not unate in x_2.

The next two propositions show that if a cover of a unate function consists only of prime cubes, then it must be a unate cover.

Proposition 3.3.2. A logic function f is monotone increasing (decreasing) in x_j if and only if no prime implicant of f has a 0 (1) in the jth position.

Proof: For the sake of simplicity, assume that f is a single-output logic function.

If part: If no prime implicant has a 0 in the jth position, any prime cover $\mathscr{C}(f)$ is unate in x_j. By Proposition 3.3.1, f must be unate in x_j.

Only if part: Assume that there is a prime c^i of f that has a 0 in the jth position. Since c^i is prime, the cube $\tilde{c}^i$ obtained by replacing the 0 in the jth position with a 1, contains at least one vertex in $\bar{f}$. For the input combination, v, corresponding to that vertex, f must be 0. Now, if we change x_j from 1 to 0 in v, the value of f changes from 0 to 1, so f is not monotone increasing. ■

It is trivial to generalize this proposition to the following.

Proposition 3.3.3. A prime cover of a unate function is unate.

Note that in the example, F was not a prime cover. A prime cover obtained from F is

$$F' = \begin{bmatrix} 1 & 2 & 0 & 4 \\ 2 & 0 & 2 & 4 \end{bmatrix}$$

which is obviously unate.

Single-output unate functions have very important properties summarized in the following propositions. These propositions could be extended to the multiple output case, but for the purpose of their use in this book the single-output results are sufficient. In what follows, we make no distinction between a cover and the completely specified logic function it induces.

Proposition 3.3.4. A unate cover is a tautology if and only if it contains a row of all 2's.

Proof of Only if Part: Assume without loss of generality that the cover represents a monotone increasing Boolean function, and hence consists solely of 1's and 2's. Note that unless the cover contains a row of all 2's, the minterm (0,0,...,0) is disjoint from every cube in the cover, and so is not contained in the union of the vertices of the cover.

The if part is trivially true. ■

Proposition 3.3.5. Let $\mathscr{C}$ be a unate cover, $\mathscr{S}$ any subset of the cubes of $\mathscr{C}$ and c any cube of $\mathscr{C}$. Then $c \subseteq \mathscr{S}$ if and only if $c \subseteq s$, for some $s \in \mathscr{S}$.

Proof of Only if Part: Suppose that $\mathscr{C}$ is a positive unate cover. Since $\mathscr{C}$ is positive unate, every cube in $\mathscr{C}$ contains the vertex (1,1,...,1) and so $c \cap s \neq \phi$, $\forall s \in \mathscr{S}$. Now suppose $c \subseteq \mathscr{S}$. This implies that $\mathscr{S}_c \equiv 1$. But by Proposition 3.3.4, a unate cover is a tautology if and only if it contains a row of all 2's. Thus there exists $s \in \mathscr{S}$, which generated this row. Thus $c \subseteq s$ for this $s \in \mathscr{S}$, which proves the only if part.

The if part is trivially true. ■

Proposition 3.3.6. Every prime of a unate function is essential.

Proof: Let $\mathscr{P}$ be the set of all primes of F (a cover of a unate function *f*). Suppose there exists $p \in \mathscr{P}$ which is not essential. Then by Proposition 2.3.1, there exists $\mathscr{S} \subseteq \mathscr{P}$, $p \notin \mathscr{S}$ such that $p \subseteq \mathscr{S}$. By Proposition 3.3.5, there exists $s \in \mathscr{S}$, such that $p \subseteq s$, contradicting the hypothesis that $p \in \mathscr{P}$. ■

Proposition 3.3.7. Let $\mathscr{C}$ be a unate cover, and $\mathscr{P}$ be the set of all primes of $b(\mathscr{C})$, the logic function specified by $\mathscr{C}$. Then $\mathscr{P} \subseteq \mathscr{C}$. If, in addition, $\mathscr{C}$ is minimal with respect to single cube containment, then $\mathscr{C} = \mathscr{P}$, and $\mathscr{C}$ is the unique minimum cover.

Proof: To prove the first part, let p be a prime of $b(\mathscr{C})$. Then $p \subseteq \mathscr{C}$, so $p \subseteq c \in \mathscr{C}$ for some cube c (by Proposition 3.3.5). But p is prime, so $p = c$. Therefore all primes occur in $\mathscr{C}$.

Now assume $\mathscr{C}$ is minimal with respect to containment. Then any $c \in \mathscr{C}$ is contained in some prime p, but we have just seen $p \in \mathscr{C}$;

therefore c = p. Hence all elements of $\mathscr{C}$ are prime, and all primes occur in $\mathscr{C}$, so $\mathscr{C} = \mathscr{P}$. ■

Proposition 3.3.7 implies that we can obtain the minimum cardinality prime cover of a unate function easily, starting from any cover, by (1) expanding each cube to a prime and (2) removing any cube which is contained in any other cube of the cover. Obviously this procedure generates a prime cover. Since the minimum-cardinality cover contains all primes, any prime cover has minimum cardinality and the procedure is correct.

The following propositions are important for constructing an efficient algorithm for finding the complement of a logic function. The proofs are straightforward and are omitted.

Proposition 3.3.8. If a logic function f is monotone increasing in x_j, then the complement of f, $\bar{f}$, is monotone increasing in $\bar{x}_j$.

Proposition 3.3.9. The complement of a unate function is unate.

Proposition 3.3.10. The cofactors of a unate function f with respect to x^j and $\bar{x}^j$ are unate.

3.4 The Choice of the Splitting Variable.

The choice of the variable to split is perhaps the most important step in the passage from the paradigm to an efficient algorithm. The choice is guided by the heuristic of making the covers $\mathscr{C}_x$ and $\mathscr{C}_{\bar{x}}$ unate after a minimum number of splittings. Since we may not be dealing with prime implicants at each stage, the logic functions f_x and $f_{\bar{x}}$ covered by $\mathscr{C}_x$ and $\mathscr{C}_{\bar{x}}$ may be unate even though the covers **are not unate.** If a cover is unate in a variable x_j, we say that x_j is a unate variable. A nonunate variable is called **binate.** The splitting variable is chosen among the binate variables: we shall choose the "**most**" binate variable in an attempt at keeping the total number of cubes which are part of both $\mathscr{C}_x$ and $\mathscr{C}_{\bar{x}}$ as small as possible. This leads to the procedure BINATE__SELECT shown in Figure 3.4.1.

```
Procedure BINATE_SELECT(𝒞)

/* Given a cover 𝒞 = {𝒞^k}, selects the "most" binate variable
/* x_ĵ for splitting. The number of variables is n.

    Begin
    for (j = 1,..., n)
        Begin
        p_0(j) ← | {c^i ∈ 𝒞 | c^i_j = 0} |           /* Count the number of
                                                    /* cubes with a 0
                                                    /* in the j^th input position.
        p_1(j) ← | {c^i ∈ 𝒞 | c^i_j = 1} |           /* Count the number of
                                                    /* cubes with a 1
                                                    /* in the j^th input position.
        End
    if ( max_j  min {p_0(j), p_1(j)} = 0) return (U ← True, 0)   /* U = True indicates 𝒞 was
                                                    /* unate and no ĵ was chosen.
    else
        Begin
        J ← {j | min (p_0(j), p_1(j))>0}
        ĵ ← argmax_{j∈J} {p_0(j) + p_1(j)}             /* Select ĵ in the set of maximizers
                                                    /* of p_0(j) + p_1(j), i.e. the
        return (U ← False; ĵ)                        /* most binate variable.
        End
    End
```

Figure 3.4.1

We now describe how to perform complementation efficiently on unate covers.

3.5 Unate Complementation

The key to a fast unate complementer is the following result.

Proposition 3.5.1. The complement of a unate cover F can be expressed as

$$\overline{F} = \overline{x}^j \overline{F}_{\overline{x}^j} + \overline{F}_{x^j} \qquad (3.5.1)$$

if F is monotone increasing in x_j, or as

$$\overline{F} = x^j\overline{F}_{x^j} + \overline{F}_{\overline{x}^j}, \tag{3.5.2}$$

if F is monotone decreasing in x_j.

Proof: We prove (3.5.1) since (3.5.2) can be obtained trivially by exchanging $\overline{x}^j$ and x^j. Since F is monotone increasing in x_j, every cube of F has either a 1 or a 2 in the jth position. Hence $F_{\overline{x}^j} \subseteq F_{x^j}$ and F can be written as

$$F = x^jF_{x^j} + F_{\overline{x}^j} = (x^j + F_{\overline{x}^j})\, F_{x^j}. \tag{3.5.3}$$

Thus we obtain (3.5.1) by complementing (3.5.3). ■

The proposition allows us to decompose the problem of computing the complement of a unate cover into two subproblems:

1. Compute $\overline{x}^j\overline{F}_{\overline{x}^j}$; this involves complementation of a reduced number of cubes (assuming F depended on x_j) and one less variable, x_j, since $|F_{\overline{x}^j}| < |F|$, and $F_{\overline{x}^j}$ contains a column of all 2's in the jth position.

2. Compute $\overline{F}_{x^j}$; this involves complementation of the original cover dropping one variable, since $|F_{x^j}| = |F|$, because F is monotone increasing in x_j.

The complementation process then is simply to choose a variable, recursively complement a cover consisting of cubes not containing that variable, and complement the original cover with that variable dropped, tack the variable onto cubes of the first result, and concatenate the two results. The recursion has a natural termination since the complement of a cover with no cubes is the universe and the complement of a cover with no variables is empty.

In the actual implementation of the algorithm, we deal with a multiple output function by complementing one output at a time. Experimentally, this has been proven to be faster and easier to execute than the multiple-output complementation.

We also use a simplified matrix representation M of the unate cover F. Let F be the matrix representation of a single-output unate cover with k cubes and n input variables. The matrix M is a Boolean matrix defined as follows.

$$M_{ij} = \begin{cases} 1 & \text{if } F_{ij} = 0 \text{ or } 1 \\ 0 & \text{if } F_{ij} = 2 \end{cases} \qquad i = 1,\ldots,k;\ j = 1,\ldots,n.$$

Note that the output part is dropped since we are dealing with a single output function only and this information is clearly redundant. A Boolean matrix requires less storage, and the operations we perform can be executed faster than if we had a matrix with integer entries. We can recover the original matrix representation by storing the information on whether F increases or decreases monotonically with respect to each x_j.

The algorithm for the complementation of unate functions is described in the three procedures of Figures 3.5.1 (UNATE_COMPLEMENT), 3.5.2 (PERS_UNATE_ COMPLEMENT), and 3.5.3 (UCOMP_SELECT).

The key to the efficiency of PERS_UNATE_COMPLEMENT is the selection of the splitting variable. To justify our selection procedure, consider the computation dictated by the successive selection of k variables, $x_1,\ldots, x_k$. The first decomposition produces

$$\overline{x}^1 \overline{F}_{\overline{x}^1} + \overline{F}_{x^1}$$

and PERS_UNATE_COMP is called to compute $\overline{F}_{x^1}$ (the first branch of the recursion). Then we obtain

$$\overline{x}^1 \overline{F}_{\overline{x}^1} + \overline{x}^2 \overline{F}_{x^1 \overline{x}^2} + \overline{F}_{x^1 x^2}$$

and the procedure is called to compute $\overline{F}_{x^1 x^2}$, to yield

$$\overline{x}^1 \overline{F}_{\overline{x}^1} + \overline{x}^2 \overline{F}_{x^1 \overline{x}^2} + \overline{x}^3 \overline{F}_{x^1 x^2 \overline{x}^3} + \overline{F}_{x^1 x^2 x^3}$$

```
Procedure UNATE__COMPLEMENT(F)

/* Given F, a matrix representation of a unate function.
/* Computes R, a matrix representation of the complement of F.
/* F has k rows and n columns.

    Begin
    M←M(F)                                /* Compute the personality matrix M of F.
    V←MONOTONE(F)                         /* V records if F is monotone increasing
                                          /* or decreasing with respect to each of the
                                          /* variables x_j.
    M̄←PERS__UNATE__COMPLEMENT(M)
    R←TRANSLATE(M̄, V)                    /* The matrix representation of the
                                          /* complement is obtained from M̄ and V.

    End
```

Figure 3.5.1

If we do not encounter special cases, at the kth stage we have

$$\bar{x}^1\bar{F}_{\bar{x}^1} + \bar{x}^2\bar{F}_{x^1\bar{x}^2} + \dots + \bar{x}^k\bar{F}_{x^1x^2\dots\bar{x}^k} + \bar{F}_{x^1\dots x^k}.$$

If there is a cube $c \in F$ such that c^i, $i = 1,\dots, k$, are the only entries different from 2, then by definition of cofactor $F_{x^1\dots x^k}$ is a tautology and the complement is obviously empty. Thus, the recursion must stop at the kth stage and we have

$$\bar{x}^1\bar{F}_{\bar{x}^1} + \bar{x}^2\bar{F}_{x^1\bar{x}^2} + \dots + \bar{x}^k\bar{F}_{x^1x^2\dots\bar{x}^k}.$$

Note that at each level of the recursion, $F_{x^1x^2\dots x^j}$ is computed from $F_{x^1x^2\dots x^{j-1}}$ by simply setting its jth column to "2". Similarly, $F_{x^1x^2\dots x^{j-1}\bar{x}^j}$ is computed by restricting to those rows of $F_{x^1x^2\dots x^{j-1}}$ which are already "2" in x_j. To speed up the process, we try to keep the level of the recursion as shallow as possible and to make the computation involved in each of the "sub complements" as simple as possible.

Since in the worst case, the recursion reaches its deepest level when all the variables of at least one cube have been split, we decide to

Procedure PERS__UNATE__COMPLEMENT(M)

/* Given M, the personality of a matrix of a unate function,
/* computes $\overline{M}$, the personality of a matrix representation of the complement.
/* M has k rows and n columns.

Begin

$\overline{M} \leftarrow \phi$	/* Initialize
$(\overline{M}, T) \leftarrow$ SPECIAL__CASES(M)	/* Computes $\overline{M}$ for special cases.
if (T = 1) **return** ($\overline{M}$)	/* $\overline{M}$ has been computed by SPECIAL__CASES
$\hat{j} \leftarrow$ UCOMP__SELECT(M)	/* Select the splitting variable
$(M^1, M^0) \leftarrow$ PERS__COFACTORS(M, $\hat{j}$)	/* Computes the personality matrix of /* the cofactors with respect to $x_{\hat{j}}$.
$\overline{M}^1 \leftarrow$ PERS__UNATE__COMPLEMENT(M^1)	/* One branch of the recursion.
$\overline{M}^0 \leftarrow$ PERS__UNATE__COMPLEMENT(M^0)	/* The other branch of the recursion.
return ($\overline{M} \leftarrow$ MERGE($\overline{M}^1$, $\overline{M}^0$))	/* The merging process essentially /* concatenates the rows of $\overline{M}^1$ with the /* rows of $\overline{M}^0$.

End

/* SPECIAL__CASES	Result	Return
/*		
/* There is a row of all 0s	The function is a tautology and	$T \leftarrow 1$, $\overline{M} \leftarrow \phi$.
/*	the complement of the function	
/*	is empty.	
/* M is empty.	The complement is a tautology.	$T \leftarrow 1$, $\overline{M} \leftarrow$[0...
/* M has only one term.	The complement is computed	$T \leftarrow 1$, and
/*	by DeMorgan's law. $\overline{M}$ has one row.	$\overline{M} \leftarrow$DEMORG
/* None of the above.	Return indication of this.	$T \leftarrow 0$, $\overline{M} \leftarrow \phi$.

Figure 3.5.2

```
Procedure UCOMP_SELECT(M)

/* Given a personality matrix M with n columns
/* and k rows, selects a splitting variable x_ĵ

Begin
   î ← SELECT1(argmin_i Σ_{j=1}^{n} M_ij)    /* Select the largest cube.
   J ← {j | M_îj = 1}                        /* Select the set of variables in
                                             /* the largest cube.
   return (ĵ ← argmax_{j∈J} Σ_{i=1}^{m} M_ij)  /* Select the most frequently
                                             /* appearing variable.
End
```

Figure 3.5.3

identify the splitting variables by choosing the variables of the largest cube in F, i.e. the cube with the maximum number of 2's. Having identified the splitting variables, we have to decide the order in which these variables are processed. This order is selected by choosing first the variables that appear most often in the other cubes of F. This has the effect of eliminating the most cubes in one of the branches of the recursion. Keeping the number of arguments (and hence the results) small has several nice effects. Besides speeding up the process, small results allow very little containment to be produced in the MERGE process. This minimizes the time wasted in eliminating unnecessary terms. This selection of the splitting variable is implemented by UCOMP_SELECT shown in Figure 3.5.3.

This strategy has proven to be so successful that the containment part of the MERGE process has been completely eliminated, so that only identification of identical cubes is done in MERGE.

We illustrate how UNATE_COMPLEMENT (Figure 3.5.1) works on the following example. Let the matrix representation of f be

$$F = \begin{matrix} 0 & 1 & 2 & 2 \\ 0 & 2 & 2 & 2 \\ 2 & 1 & 2 & 1 \\ 2 & 1 & 0 & 2 \\ 2 & 1 & 0 & 1 \end{matrix}$$

Note that F is unate. The Boolean matrix M associated with F is

$$M = \begin{matrix} 1 & 1 & 0 & 0 \\ 1 & 0 & 0 & 0 \\ 0 & 1 & 0 & 1 \\ 0 & 1 & 1 & 0 \\ 0 & 1 & 1 & 1 \end{matrix}$$

Applying UNATE_COMPLEMENT to this example, we obtain, from results at the leaves of the recursion tree,

$$\overline{M} = \begin{bmatrix} 1 & 1 & 0 & 0 \\ 1 & 0 & 1 & 1 \end{bmatrix},$$

corresponding to

$$\overline{F} = \begin{bmatrix} 1 & 0 & 2 & 2 \\ 1 & 2 & 1 & 0 \end{bmatrix},$$

since now a 1 in $\overline{M}$ represents the complement of the corresponding variable in F.

It is useful to note that unate complementation is intimately related to the concept of a column cover of a binary matrix, a concept which is part of IRREDUNDANT_COVER and EXPAND discussed in Sections 4.5 and 4.3 respectively. A "column cover" of a binary matrix is defined as a set of columns L such that

$$\sum_{j \in L} M_{ij} \geq 1, \ \forall i.$$

Proposition 3.5.2. Each row i of $\overline{M}$, the binary matrix complement of M, corresponds to a column cover L of M where $j \in L$ if and only if $\overline{M}_{ij} = 1$. The rows of $\overline{M}$ include the set of all minimal column covers of M. (If $\overline{M}$ was minimal with respect to containment, then $\overline{M}$ would be precisely the set of all minimal column covers of M.)

The proof (omitted) follows from the fact that each cube of $\overline{F}$ must be orthogonal to all the cubes of F. This translates directly into the statement that each row of $\overline{M}$ is a column cover of M.

3.6 Simplify

We now illustrate the use of the unate recursive paradigm to construct a heuristic minimization algorithm, SIMPLIFY. Even though this is not one of the algorithms of ESPRESSO-II, we discuss it for illustrative purposes, and also because it has proven to be an extremely valuable algorithm in situations where fast but good single-output minimization is required. SIMPLIFY is designed for speed which is achieved by sacrificing guarantees that each cube of the cover be prime and irredundant. In addition, SIMPLIFY will be described for single-output functions only. A multiple-output version of the algorithm was implemented for PLA minimization, but the results on practical problems were disappointing. Since the single-output version performs so well, we conjecture that the disappointing results may be due to the way the output part is handled during the merging process. However, this must be left for future research. SIMPLIFY is also restricted by necessity to apply to completely specified functions, i.e. where the don't-care set is null.

SIMPLIFY is very similar to the single-output complementer, COMP1 which will be described in Section 4.1. The basic flow is exactly the same and one algorithm can be transformed into the other by altering the leaf conditions, i.e. altering the computations that are done when the cover is unate or when a special case applies. Indeed when the leaf conditions are made specific to the complement operation we have COMP1 and when made specific to the minimization operation we have SIMPLIFY. Since COMP1 is an integral part of ESPRESSO-II, we will delay its description until Section 4.1.

The unate recursive paradigm is applied in the following way. Expressing f in the Shannon expansion $f = x_j f_{x_j} + \overline{x}_j f_{\overline{x}_j}$, we obtain simplification by applying the merging and simplification operations to the cofactors, to obtain the recursive formula

$$\text{SIMPLIFY}(f) = \text{MWC}(x_j \ \text{SIMPLIFY}(f_{x_j}) + \bar{x}_j \ \text{SIMPLIFY}(f_{\bar{x}_j}))$$

where MWC = MERGE_WITH_CONTAINMENT.

We have already discussed in Proposition 3.3.7 how one minimizes a single-output unate cover by merely eliminating each cube that is contained in any other cube of the cover. At this point we are guaranteed that the cover thus constructed is the unique **minimum** prime cover for this unate function.

It is in the merging process that primeness and irredundancy is sacrificed. In principle, one can preserve primality and irredundancy at each merge step; however, to do so, an expensive calculation is required. Experience has shown that by using MERGE_WITH_CONTAINMENT, we obtain efficiency without giving up much in quality. Much of the time, this kind of merging produces the optimum result. Another key to the quality of the results of SIMPLIFY, is in choosing the splitting variable at each stage, so that the recursion will not get too deep before a unate leaf is encountered. In that case, there are only a few mergings required to get to the final result. Since the unate leaf results are optimum, the relatively few mergings should not displace the final result too far from the optimum.

The procedure SIMPLIFY is shown in Figure 3.6.1. Notice the similarity with COMP1 in Figure 4.1.2. UNATE_SIMPLIFY is a procedure which eliminates any term of a cover contained in any other term. By Proposition 3.3.7 UNATE_SIMPLIFY could be called CONTAIN. Merging is done with MERGE_WITH_CONTAINMENT described in Section 3.2. The variable that is chosen for splitting in the Shannon expansion is selected by BINATE_SELECT as the variable which is "most binate". This heuristic keeps the depth of recursion small in most cases. Finally, the new cover is tested so that the returned cover $\mathcal{F}'$ is not larger than the original one $\mathcal{F}$.

We illustrate the SIMPLIFY algorithm first with an easy exam-

```
Procedure SIMPLIFY(𝓕)

/* Given a cover 𝓕 of f, a single-output logic function,
/* returns 𝓕' a smaller or equal sized cover.

   Begin
   if (𝓕 unate) return (𝓕' ← UNATE_SIMPLIFY(𝓕))  /* UNATE_SIMPLIFY, equals
                                                 /* containment removal.
   j ← BINATE_SELECT(𝓕)                          /* Select the most binate variable.

   𝓕' ← MERGE_WITH_CONTAINMENT
        (x_j SIMPLIFY(𝓕_{x_j}), x̄_j SIMPLIFY(𝓕_{x̄_j}))
   if (|𝓕| < |𝓕'|) return (𝓕' ← 𝓕)               /* Don't let the cover increase.
   else return (𝓕')
   End
```

Figure 3.6.1

ple. Consider the three cube cover given in matrix form by

$$f = \begin{matrix} x_1 & x_2 & x_3 \\ 0 & 2 & 2 \\ 1 & 1 & 2 \\ 2 & 1 & 1 \end{matrix}$$

2

The only binate variable is x_1. We form the cofactor

$$f_{x_1} = \begin{bmatrix} 2 & 1 & 2 \\ 2 & 1 & 1 \end{bmatrix}$$

and discover it is unate. Observing that the second cube is contained in the first, we simplify it to

$$f_{x_1} = [2 \ \ 1 \ \ 2].$$

The other cofactor is

$$f_{\bar{x}_1} = \begin{bmatrix} 2 & 2 & 2 \\ 2 & 1 & 1 \end{bmatrix}.$$

which simplifies to $f_{\bar{x}_1} = [2\ 2\ 2]$. Now we perform MERGE_WITH_CONTAINMENT. Since $f_{x_1} \subseteq f_{\bar{x}_1}$, we express f as

$f_{x_1} + \bar{x}_1 f_{\bar{x}_1}$ (instead of in the more obvious but less simplified form $x_1 f_{x_1} + \bar{x}_1 f_{\bar{x}_1}$), and obtain

$$f = \begin{bmatrix} 2 & 1 & 2 \\ 0 & 2 & 2 \end{bmatrix}$$

as the final result. Clearly, the minimum irredundant cover has been obtained, since the final f is unate.

As a more complicated example, consider the function

$$\begin{aligned} f = {} & \bar{B}\bar{C}\bar{D}\bar{F} + \bar{A}\bar{B}CE + ACF + \bar{A}BD + B\bar{E}F \\ & + B\bar{C}\bar{D}F + A\bar{B}C\bar{D}\bar{E} + \bar{B}\bar{C}\bar{D}F + A\bar{B}C\bar{D}E + \\ & B\bar{C}\bar{D}F + ABD\bar{F} + ABDF, \end{aligned}$$

which in matrix form is represented by

$$f = \begin{matrix} 2 & 0 & 0 & 0 & 2 & 0 \\ 0 & 0 & 1 & 2 & 1 & 2 \\ 1 & 2 & 1 & 2 & 2 & 1 \\ 0 & 1 & 2 & 1 & 2 & 2 \\ 2 & 1 & 2 & 2 & 0 & 1 \\ 2 & 1 & 0 & 0 & 2 & 1 \\ 1 & 0 & 1 & 0 & 0 & 2 \\ 2 & 0 & 0 & 0 & 2 & 1 \\ 1 & 0 & 1 & 0 & 1 & 2 \\ 2 & 1 & 0 & 0 & 2 & 0 \\ 1 & 1 & 2 & 1 & 2 & 0 \\ 1 & 1 & 2 & 1 & 2 & 1 \end{matrix}$$

The most binate variable is B. The cofactor of f with respect to B is

$$f_B = \begin{matrix} 1 & 2 & 1 & 2 & 2 & 1 \\ 0 & 2 & 2 & 1 & 2 & 2 \\ 2 & 2 & 2 & 2 & 0 & 1 \\ 2 & 2 & 0 & 0 & 2 & 1 \\ 2 & 2 & 0 & 0 & 2 & 0 \\ 1 & 2 & 2 & 1 & 2 & 0 \\ 1 & 2 & 2 & 1 & 2 & 1 \end{matrix}$$

This simplifies (we omit the details) to

$$f_B = \begin{matrix} 1 & 2 & 2 & 2 & 2 & 1 \\ 2 & 2 & 2 & 1 & 2 & 2 \\ 2 & 2 & 2 & 2 & 0 & 1 \\ 2 & 2 & 0 & 2 & 2 & 2 \end{matrix}$$

Similarly, the simplified cofactor with respect to $\overline{B}$ is

$$f_{\overline{B}} = \begin{matrix} 1 & 2 & 1 & 2 & 2 & 1 \\ 0 & 2 & 1 & 2 & 1 & 2 \\ 2 & 2 & 0 & 0 & 2 & 2 \\ 1 & 2 & 2 & 0 & 2 & 2 \end{matrix}$$

Now note that rows 1 and 4 of f_B contain rows 1 and 3 of $f_{\overline{B}}$ respectively. MERGE_WITH_CONTAINMENT then yields

$$f = \begin{matrix} 1 & 1 & 2 & 2 & 2 & 1 \\ 2 & 1 & 2 & 1 & 2 & 2 \\ 2 & 1 & 2 & 2 & 0 & 1 \\ 2 & 1 & 0 & 2 & 2 & 2 \\ 0 & 0 & 1 & 2 & 1 & 2 \\ 1 & 0 & 2 & 0 & 2 & 2 \\ 1 & 2 & 1 & 2 & 2 & 1 \\ 2 & 2 & 0 & 0 & 2 & 2 \end{matrix},$$

i.e.,

$$f = ABF + BD + B\overline{E}F + B\overline{C} + \overline{A}\overline{B}CE + A\overline{B}\overline{D} + ACF + \overline{C}\overline{D}$$

However, using a more powerful minimizer, we can obtain

$$f = BD + B\overline{E}F + \overline{A}\overline{B}CE + A\overline{B}\overline{D} + ACF + \overline{C}\overline{D}$$

Using SIMPLIFY we obtained a redundant cover, since $B\overline{C}$ is covered by $BD + \overline{C}\overline{D}$ and ABF is covered by $ACF + \overline{C}\overline{D} + BD$. However, we did obtain a prime cover in this case.

Chapter 4

THE ESPRESSO-II MINIMIZATION LOOP AND ALGORITHMS

4.0 Introduction

ESPRESSO-II receives as its inputs $\mathcal{F}$ and $\mathcal{D}$, cube covers of the on-set and the don't-care set of an incompletely specified Boolean function ff. Optionally, it can accept input $\mathcal{F}$ and $\mathcal{R}$, cube covers of the on- and off-sets. It returns as its output a "minimized" cover. As discussed in Chapter 1, the objectives of ESPRESSO-II are to minimize:

NPT: the number of product terms in the cover;

NLI: the number of literals (non-2's) in the input parts of the cover;

NLO: the number of literals in the output parts.

The ESPRESSO-II minimization procedure defines a vector objective function

$$\Phi = (\text{NPT, NLI, NLO})$$

and continues to iterate through its main minimization loop until none of the 3 components of Φ have been reduced since the last pass through the loop. The procedure is illustrated in Figure 4.0.1.

Before beginning the minimization process, the preprocessor UNWRAP is first applied to the incoming data. This procedure unravels any output sharing that may be present in the incoming data; each incoming cube which feeds k different outputs (has k 4's in the output part) is replaced by k cubes each feeding one output. Although the resulting cover is less optimal, it gives a less biased starting point for subsequent minimizations; in particular, EXPAND can then decide what output sharing is desirable. This procedure, suggested by R. Rudell, leads to more reliable and input-independent results.

As described in Chapter 1, ESPRESSO-II minimization involves seven basic routines: COMPLEMENT, EXPAND, ESSENTIAL__PRIMES, IRREDUNDANT__COVER, REDUCE, LAST__GASP, and MAKE__SPARSE. In addition, the Boolean processing performed by many of these routines relies heavily on the TAUTOLOGY algorithm, which is used to determine when one cube is covered by another set of cubes. In this chapter, there are eight subsections which treat these routines in turn, and a ninth which describes a partitioning procedure for large problems. Here, we will describe how these routines fit together to form ESPRESSO-II.

COMPLEMENT computes $\mathscr{R}$ (the offset of $\mathscr{F}$), or $\mathscr{D}$ if both $\mathscr{F}$ and $\mathscr{R}$ are given as input. The complement plays an essential role in the subsequent processing, because it can be used to quickly determine whether or not a given cube is an implicant. This facilitates the selection of prime implicants in the EXPAND procedure.

EXPAND replaces the cubes of $\mathscr{F}$ by prime implicants and assures the cover is minimal with respect to single-cube containment. Thus EXPAND reduces the number of cubes in $\mathscr{F}$ and the number of cares in the input part of $\mathscr{F}$, improving two of our objectives.

ESSENTIAL__PRIMES locates those prime implicants which must appear in every cover of $\mathscr{F}$. (These are called essential primes).

```
Procedure ESPRESSO-II (𝓕, 𝓓)

/* Given 𝓕, a cover of ff = {f,d,r} = (on-set, don't-care, off-set)
/* and 𝓓 a cover of d, minimizes Φ(𝓕) = (NPT, NLI, NLO)
/* where NPT is the number of cubes
/* NLI is the number of input literals and
/* NLO is the number of output literals.
/* Returns a minimized cover 𝓕 and Φ its cost.

    Begin
    𝓕 ← UNWRAP(𝓕)                              /* Preprocess 𝓕
    𝓡 ← COMPLEMENT(𝓕, 𝓓)
    Φ1* ← Φ2* ← Φ3* ← Φ4* ← COST(𝓕)            /* Initialize Cost

    LOOP1: (Φ,𝓕) ← EXPAND(𝓕, 𝓡)                /* 𝓕 is prime and SCC-minimal.

        if (First-Pass)                         /* Move essential primes into
        (Φ,𝓕,𝓓,𝓔) ← ESSENTIAL__PRIMES(𝓕,𝓓)    /* don't care set.

        if (Φ ≡ Φ1*) goto OUT                   /* Check termination criterion.
        Φ1* ← Φ
        (Φ,𝓕) ← IRREDUNDANT__COVER(𝓕,𝓓)        /* 𝓕 is now a minimal cover
        if (Φ ≡ Φ2*) goto OUT                   /* of prime implicants.
        Φ2* ← Φ
    LOOP2: (Φ,𝓕) ← REDUCE (𝓕,𝓓)                /* Each cube of 𝓕 replaced by
                                                /* smallest cube containing its
                                                /* "essential" vertices.

        if (Φ ≡ Φ3*) goto OUT
        Φ3* ← Φ
        goto LOOP1

    OUT: if (Φ ≡ Φ4*) goto QUIT
        (Φ',𝓕) ← LAST__GASP (𝓕,𝓓,𝓡)            /*If no further improvement,
        if (Φ ≡ Φ') goto QUIT                   /*terminate.
        Φ1* ← Φ2* ← Φ3* ← Φ4* ← Φ'
        goto LOOP2

    QUIT: 𝓕 ← 𝓕 ∪ 𝓔                            /* Put essential primes 𝓔 back
        𝓓 ← 𝓓 − 𝓔                               /* into cover and take them out of 𝓓.
        (Φ,𝓕) ← MAKE__SPARSE (𝓕,𝓓,𝓡)           /* Concentrate on literals.
    return (Φ, 𝓕)
    End
```

Figure 4.0.1. ESPRESSO-II Minimization Procedure.

Since these primes must appear, once they are identified they are removed from $\mathcal{F}$ and added to the don't-care set $\mathcal{D}$. This prevents the essential prime cubes from being processed needlessly during the minimization loop. ESSENTIAL_PRIMES is executed only during the first pass.

Procedure IRREDUNDANT_COVER sorts the cover $\mathcal{F}$ into relatively essential, partially redundant and totally redundant subcovers. The totally redundant cubes are discarded, and a minimal subset of the partially redundant subcover is retained which, in concert with the relatively essential subcover and $\mathcal{D}$, suffices to cover all the minterms of $\mathcal{F}$. Thus after the call to IRREDUNDANT_COVER, $\mathcal{F}$ is a minimal cover for $\mathcal{f}$.

REDUCE facilitates improvement over the local minimum obtained by IRREDUNDANT_COVER. REDUCE takes, in turn, each cube $c \in \mathcal{F}$, and reduces it to the smallest cube $\underline{c}$ containing all minterms of c not contained by $(\mathcal{F}-\{c\}) \cup \mathcal{D}$; then sets $\mathcal{F} = (\mathcal{F}-\{c\}) \cup \{\underline{c}\}$. This typically produces a nonprime cover, in which many of the original cubes have contracted to a smaller size. The benefit, realized when these cubes are re-expanded, is twofold. First, a smaller cube can generally be expanded in more directions than a larger cube. Secondly, smaller cubes have more chance of being covered by the expansions of other cubes. In this way REDUCE often allows us to move away from a locally minimal solution towards a yet smaller cover. Since REDUCE contracts cubes in the cover sequentially, its result depends upon an ordering heuristic.

LAST_GASP is reminiscent of REDUCE but uses an order independent reduction process. For each cube c^i in the cover $\mathcal{F}$, we compute $\underline{c}^i$, the smallest cube containing the portion of c not covered by $(\mathcal{F}-\{c^i\}) \cup \mathcal{D}$. Since the cubes are reduced individually rather than in sequence, the result does not depend upon the ordering. Note that the cubes $\{\underline{c}^i\}$ do not necessarily form a cover of $\mathcal{f}$. We then use a variation of EXPAND to attempt to find at least one prime which covers two of the reduced cubes $\{\underline{c}^i\}$. The prime or primes with this property are collected into a cover $\mathcal{H}$ and $\mathcal{F}$ is replaced by IRREDUNDANT_COVER $(\mathcal{F} \cup \mathcal{H})$. When $\mathcal{H}$ is nonempty, we have always observed a decrease in $|\mathcal{F}|$ as a result of this procedure.

ESPRESSO-II processing concludes with the procedure MAKE_SPARSE. The essential primes are first taken out of the don't-care set $\mathscr{D}$ and put back into the cover $\mathscr{F}$. MAKE_SPARSE regards the number of cubes in the cover as final and attempts to reduce the total number of literals by "lowering" the outputs and "raising" the inputs. MAKE_SPARSE attempts to make the final cover minimal, in the sense that no product term, input literal, or output literal can be removed while retaining coverage of $\mathscr{f}$.

We now describe the flow of execution for ESPRESSO-II. The UNWRAP and COMPLEMENT operations are carried out to setup the covers $\mathscr{F}$, $\mathscr{R}$ and $\mathscr{D}$ for the onset, offset and don't-care set of $\mathscr{f}$. Then the four "result recorders" $\Phi 1^*$, $\Phi 2^*$, $\Phi 3^*$, $\Phi 4^*$ are each loaded with the number of terms, input literals, and output literals present in the initial cover $\mathscr{F}$. These four vectors are used to record the quality of the cover obtained after the last execution of EXPAND, IRREDUNDANT_COVER, REDUCE and LAST_GASP respectively.

The main loop executes EXPAND, followed by IREDUNDANT COVER and then REDUCE. After each step we consult the result recorders to determine if the quality of the cover has improved. If no improvement is detected, we branch to OUT, which initiates LAST_GASP. As described above, LAST_GASP searches systematically for additional primes which will be beneficial to the cover. If some are found, we re-enter the main iteration at LOOP2, to reduce the new cover prior to expansion. Otherwise, we execute MAKE_SPARSE and terminate with our final result.

In the following sections we discuss these procedures in detail.

4.1 Complementation

In this section, we describe the "multiple-output" complementer, which is essentially an iteration over each output using the single-output complementer based on the unate recursive paradigm.

As described in Section 3, we compute the complement of a given Boolean function by recursive use of the "Shannon Expansion"

$$\bar{f} \equiv x_j \bar{f}_{x_j} + \bar{x}_j \bar{f}_{\bar{x}_j}.$$

At each node of the (binary) recursion tree, the "splitting variable", x_j, is chosen so that the cofactors in the above expression become successively "more unate". The recursion continues until, at each leaf of the recursion tree, a unate function or one of the special cases described below is encountered. The special cases require only trivial computation, and serve to trim the recursion tree. If a unate function is encountered we transfer control to the unate complementation routine described in Section 3.5.

We have elected to compute and represent the complement of the given multiple-output function as the concatenation of the individual single-output functions, for the following reasons. First, our experiments have indicated that multiple-output complementation seldom saves computing time and sometimes leads to huge CPU expenditures when the usual disjoint sharp operations are employed. Also, we have implemented multiple-output unate complementation and this usually takes longer than single-output unate complementation. Another advantage of the single-output format is that it permits significant efficiency and uniformity of treatment in the the EXPAND operation (cf., Section 4.3 below). In fact, EXPAND is the only major operation in the ESPRESSO-II program which uses the complement. Of course, the single-output format does require more storage since many input product terms will often be duplicated.

The following steps are used to produce the single-output form of the complement. First the given multiple-output cover is unraveled into its single-output equivalents. The don't-care set of input cubes, which was computed in the set-up phase, is then appended to each function in turn, and finally the complement of this single function is computed. The result is then converted to binary format and appended to the overall result.

The procedure for multiple-output complementation is shown in Figure 4.1.1. EXTRACT forms the ith single-output function and its don't-care set. COMP1 is the single-output function complementer (cf. Figure 4.1.2).

The complement of a single function is obtained by first taking advantage of some special cases, and then performing the generic de-

```
Procedure COMPLEMENT(𝓕, 𝓓)

/* Given a cover 𝓕 of ff = {f,d,r} and 𝓓, a cover
/* of d, returns 𝓡, a cover of r.  𝓡 is a representation
/* of the off-sets with no input-cube sharing among the outputs.
/* m is the number of outputs

    Begin

        𝓡 ← φ
        for     (i = 1,..., m)
                Begin
                (𝓕_i,𝓓_i) ← EXTRACT(𝓕,𝓓,i)  /* Extract the single-output
                                             /* function and don't-care
                                             /* set for the i^th output.

                𝓡_i ← COMP1(𝓕_i ∪ 𝓓_i)       /* Complement the union of the
                                             /* single-output function and
                                             /* its don't-care set

                𝓡 ← 𝓡, 𝓡_i                   /* Append to other
                                             /* off-set functions
                End
        Return (𝓡)

    End
```

Figure 4.1.1 Multiple-output function complementation

composition into unate functions referred to in Section 3 as the "unate recursive paradigm".

The first special case occurs if our representation of the given function has a cube of all 2's (the universe in the given subspace), in which case the complement is empty. Of course this will never be true at the top level of our recursive complementation procedure, but may occur in any of the recursive calls when we attempt to complement a cofactor.

The second special case occurs if the cover $\mathcal{F}$ is unate i.e., if in each column of the matrix representation, there are not both 1's and 0's.

In this case the special procedure UNATE_COMPLEMENT is called and its result returned.

The next case occurs if either a column of all 0's or all 1's occurs. In this case we extract a cube c which represents the columns where this case occurs. For each j, if column j has only 0's then $c_j = 0$, if column j has only 1's then $c_j = 1$, and otherwise $c_j = 2$. Thus the cover $\mathcal{F}$ can be written as

$$\mathcal{F} = c \cap \mathcal{F}_c$$

and by De Morgan's law

$$\overline{\mathcal{F}} = \bar{c} \cup \overline{\mathcal{F}}_c.$$

Since c is unate, its complement can be obtained using UNATE_COMPLEMENT (or a trivial procedure using De Morgan's law).

Finally, we have the case where $\mathcal{F}$ is apparently binate and no cube factors out of $\mathcal{F}$. We therefore choose among the apparently binate variables (columns which have both 0's and 1's) the one which is "most binate" using the procedure BINATE_SELECT discussed in Section 3.4 (cf., Figure 3.4.1). The two factors for variable x_j are then complemented by calling the single-output complementer COMP1 again. The results are then merged using the procedure MERGE_WITH_CONTAINMENT described in Section 3.2 (cf., Figure 3.2.1). This result is then added to $\bar{c}$ to obtain the final result. The entire procedure COMP1 is shown in Figure 4.1.2.

We illustrate the above points with the following example. Suppose that a Boolean function *f* is represented by the cover

$$\mathcal{F} = \begin{array}{ccccc} x_1 & x_2 & x_3 & x_4 & x_5 \\ 2 & 1 & 1 & 1 & 1 \\ 2 & 1 & 1 & 1 & 0 \\ 2 & 1 & 0 & 1 & 2 \\ 2 & 1 & 0 & 2 & 1 \end{array}$$

First, we discover that x_2, is a column containing all 1's. Thus we have

```
Procedure COMP1 (𝓕)

/* Given a cover 𝓕 of ff = {f,d,r}, a single output logic function,
/* returns 𝓡, a cover of the offset r.
/* n is the number of input variables.

  Begin
  if (row of all 2's) Return (𝓡 ← φ)
  if (𝓕 unate) Return (𝓡 ← UNATE_COMPLEMENT(𝓕))
  c ← 𝓕^1                                               /* Initialize with first cu
  for (j = 1,..., n)                                     /* Extract cube c
      Begin                                              /* representing columns
      for (i = 2,..., |𝓕|)                               /* of all 0's or all 1's
          If (c_j ≠ 𝓕^i_j) then c_j ← 2
      End
  𝓡 ← UNATE_COMPLEMENT({c})                              /* Complement cube c
  𝓕 ← 𝓕_c

  j ← BINATE_SELECT(𝓕)                                   /* Select the most
                                                         /* binate variable.
  𝓡 ← 𝓡, MERGE_WITH_CONTAINMENT(COMP1(𝓕_xj), COMP1(𝓕_x̄j)) /* Complement cofactoı
                                                         /* and merge. Append
                                                         /* to partial result 𝓡

  Return
  End
```

Figure 4.1.2 Single-output complementation

$$f \equiv x_2 \bar{f}_{x_2}, \quad \bar{f} \equiv \bar{x}_2 + \bar{f}_{x_2}, \quad \mathcal{F}_{x_2} = \begin{matrix} 2 & 2 & 1 & 1 & 1 \\ 2 & 2 & 1 & 1 & 0 \\ 2 & 2 & 0 & 1 & 2 \\ 2 & 2 & 0 & 2 & 1 \end{matrix}$$

and it remains to compute $\bar{f}_{x_2}$. Now $\mathcal{F}_{x_2}$ does not fall into any of the special cases, so we recur by selecting the most binate variable for splitting; in this case x_3 is chosen by BINATE_SELECT. Thus we obtain

$$\bar{f}_{x_2} \equiv x_3 \bar{f}_{x_2 x_3} + \bar{x}_3 \bar{f}_{x_2 \bar{x}_3}, \quad \mathcal{F}_{x_2 x_3} = \begin{bmatrix} 2 & 2 & 2 & 1 & 1 \\ 2 & 2 & 2 & 1 & 0 \end{bmatrix},$$

$$\mathcal{F}_{x_2\bar{x}_3} = \begin{bmatrix} 2 & 2 & 2 & 1 & 2 \\ 2 & 2 & 2 & 2 & 1 \end{bmatrix}$$

$\mathcal{F}_{x_2\bar{x}_3}$ is unate and so is passed to the unate complementer. The recursion tree is completed by selecting x_5 as the next splitting variable, since each of the cofactors of $\mathcal{F}_{x_2x_3}$ thus produced are unate.

To illustrate the merging operation of the unate recursive paradigm, consider the action taken at the node of the recursion tree corresponding to the splitting and subsequent merging of variable x_3. At previous nodes in the upward traversal of the tree, suppose we have obtained the cofactor-covers

$$\overline{\mathcal{F}}_{x_3} = \begin{matrix} 22200 \\ 21211 \end{matrix}, \quad \overline{\mathcal{F}}_{\bar{x}_3} = \begin{matrix} 22200 \\ 21221 \end{matrix}.$$

In this case, both the cofactor covers contain the single cube (22200), so this single cube replaces (22000) and (22100). The second cube of the first cover is covered by the second cube of the second cover. This implies that 21211 remains unchanged and 21221 becomes 21021. Thus

$$\overline{\mathcal{F}} = \begin{matrix} 22200 \\ 21211 \\ 21021. \end{matrix}$$

In ESPRESSO-II, we are only interested in a representation of the off-set but not necessarily in a minimal representation since the only use of the off-set is to build the blocking matrix in EXPAND. Therefore, the default option used in ESPRESSO-II is to save time in COMPLEMENT by choosing to avoid the pairwise cube containment in the merging operation, and simply checking for identical cubes in the two cofactors. That is, within the COMPLEMENT procedure the flag CONTAIN is set to 0. In the above merging example, with CONTAIN set to 0, we would obtain

$$\overline{\mathcal{F}} = \begin{matrix} 22200 \\ 21111 \\ 21021. \end{matrix}$$

4.2 Tautology.

Answering the tautology question (deciding if $f \equiv 1$) is the most fundamental Boolean operation required by ESPRESSO-II. Tautology computations form an essential part of the procedures IRREDUNDANT_COVER, REDUCE, ESSENTIAL_PRIMES and LAST_GASP. Thus an efficient tautology algorithm is vital to the success of ESPRESSO-II.

In this section we begin, for simplicity, with a description of a straightforward recursive procedure for single-output tautology. We then describe several modifications to this "Vanilla tautology" algorithm which increase its practical efficiency. Finally, we discuss the complications that arise in the multiple-output case, and describe the ideas we have used to cope with them.

4.2.1 Vanilla Recursive Tautology.

A basic recursive procedure, called TAUTOLOGY, for answering the tautology question for a Boolean function, $f = b(\mathcal{F})$, is illustrated in Figure 4.2.1. The input is $F = M(\mathcal{F})$, a matrix representation of the given Boolean function f. When $\mathcal{F}$ represents a single-output Boolean function, and the calls to subprocedures UNATE_REDUCTION and COMPONENT_REDUCTION (discussed in Sections 4.2.2-4.2.4) are deleted, this procedure reduces to one which, in the light of our recent developments, we shall refer to as "Vanilla" tautology. The "Vanilla" version of TAUTOLOGY has been studied previously [MOR 70] and has been used in PRESTO [BRO 81].

Processing begins by testing for the various special cases enumerated as comments at the bottom of the procedure. The actions taken for each special case are indicated. The answer to the tautology question is trivially "yes" if there is a row of 2's in F, and trivially "no" if F has a column of all 1's or all 0's. A bound on the number of minterms of f covered by F can be obtained by counting the number of 2's per row. For example, if a cube has k 2's, then it contains exactly 2^k minterms. We sum the number of minterms contained in each cube of the cover to obtain an upper bound. If 2^n exceeds this bound, where n is the number of columns, then the answer is "no". If $n \leq 7$, then this

```
Procedure TAUTOLOGY (𝓕)

/* Given a cover 𝓕, answers Tautology Question: b(𝓕) ≡ 1?
/* Returns 1 if b(𝓕) ≡ 1 (𝓕 is tautology) else 0.
/* n = number of columns. Single - output case.

   Begin
   T←SPECIAL__CASES(𝓕,n)                              /* Returns -1 if tautology
                                                      /* question not answered.

   if(T = -1) then (T,𝓕)←UNATE__REDUCTION(𝓕)          /* Exploit Theorem 4.2.1
                                                      /* and Proposition 3.3.4.

   if(T = -1) then (T,𝓕)←COMPONENT__REDUCTION(𝓕)      /* Answers tautology question
                                                      /* if 𝓕 has components.

   if(T≠-1) return (T)                                /* Go on if no answer.
   j←BINATE__SELECT(𝓕)                                /* Most binate variable.
   if (0 = TAUTOLOGY (𝓕_xj)) return (T←0)             /* Question is answered
                                                      /* by recursive call.

   if (0 = TAUTOLOGY (𝓕_x̄j)) return (T←0)

   return (T←1)                                       /* Question is answered.
   End

/*      SPECIAL__CASES                          Result          Returned
/*
/*      1) row of 2's                        → tautology      (T = 1)
/*      2) column of all 1's or all 0's      → no tautology   (T = 0)
/*      3) deficient vertex count            → no tautology   (T = 0)
/*      4) truth table if ≤ 7 inputs or no 2's →              ?
/*      5) none of the above                 → recur          (T = -1)
```

Figure 4.2.1

minterm test can be made precise and efficient by mapping F onto a truth table.

Now assume no special cases occur. UNATE__REDUCTION and COMPONENT__REDUCTION are then skipped in "Vanilla" TAUTOLOGY. We select a "splitting variable", x_j, according to the heuristic selection rule BINATE__SELECT described in Section 3.4 (cf., Figure 3.4.1). The matrix representations $\mathcal{F}_{x^j}$ and $\mathcal{F}_{\bar{x}^j}$ of the cofactors f_{x^j} and $f_{\bar{x}^j}$ of the given Boolean function are computed. Then procedure TAUTOLOGY is recursively called with arguments

$\mathcal{F}_{x^j}$ and $\mathcal{F}_{\bar{x}^j}$. The tautology question for $\mathcal{F}$ is answered positively if and only if it is answered positively for both $\mathcal{F}_{x_j}$ and $\mathcal{F}_{\bar{x}^j}$ (cf. Proposition 3.1.3). Ultimately, the recursion stops when SPECIAL__CASES declares that a given node of the recursion tree is a leaf.

The flow of data for procedure "Vanilla" TAUTOLOGY is illustrated in Figure 4.2.2. For purposes of illustration, the early termination of the process by SPECIAL__CASES has been partially omitted. What emerges is a binary recursion tree with a matrix representing a cofactor cover as the attribute of each node. Each branch of the tree has as attribute the sequence of splitting variables which corresponds to the node-to-datum path of the tree. Each vertex is either a leaf at which the tautology question is answered or has two descendents for which the question is asked recursively. The leaf answers are always given by SPECIAL__CASES (cases 1 and 2 are illustrated in Figure 4.2.2).

The dashed line labelled UNATE__REDUCTION at the top of the figure indicates the "shortcut solution" that can be obtained by observing that certain variables (here x_2, x_3) are unate. This strategy will be described in the following section.

Notice in the example that the rows of $\mathcal{F}$ with 1 or 2 (0 or 2) in the selected splitting column reappear in the left descendent (right descendent) with the value in the splitting column replaced by 2. This is in accordance with the definition of cofactor given in Section 3.1. Note also that the columns of the matrix attribute of each node are separated by a vertical bar into already split (left) and unsplit (right) partitions. The "already split" columns are all 2's, indicating that these variables *do not really appear* in the calculations at descendant nodes in the binary tree. Finally, note that the number of terms (rows) in the matrices decreases, hopefully dramatically with the level of recursion. It is precisely on this hope that the efficiency of the Vanilla Tautology procedure relies. If there are n input variables there will be $\leq 2^n$ nodes in the recursion tree. If F has many 2's, however, entire subtrees will be removed. In the example, this occurred at the right descendent of the root, which identified the entire half space $\bar{x}^1$, as an implicant of f. Here, the tautology question is answered positively because $\mathcal{F}_{\bar{x}^j}$ contains

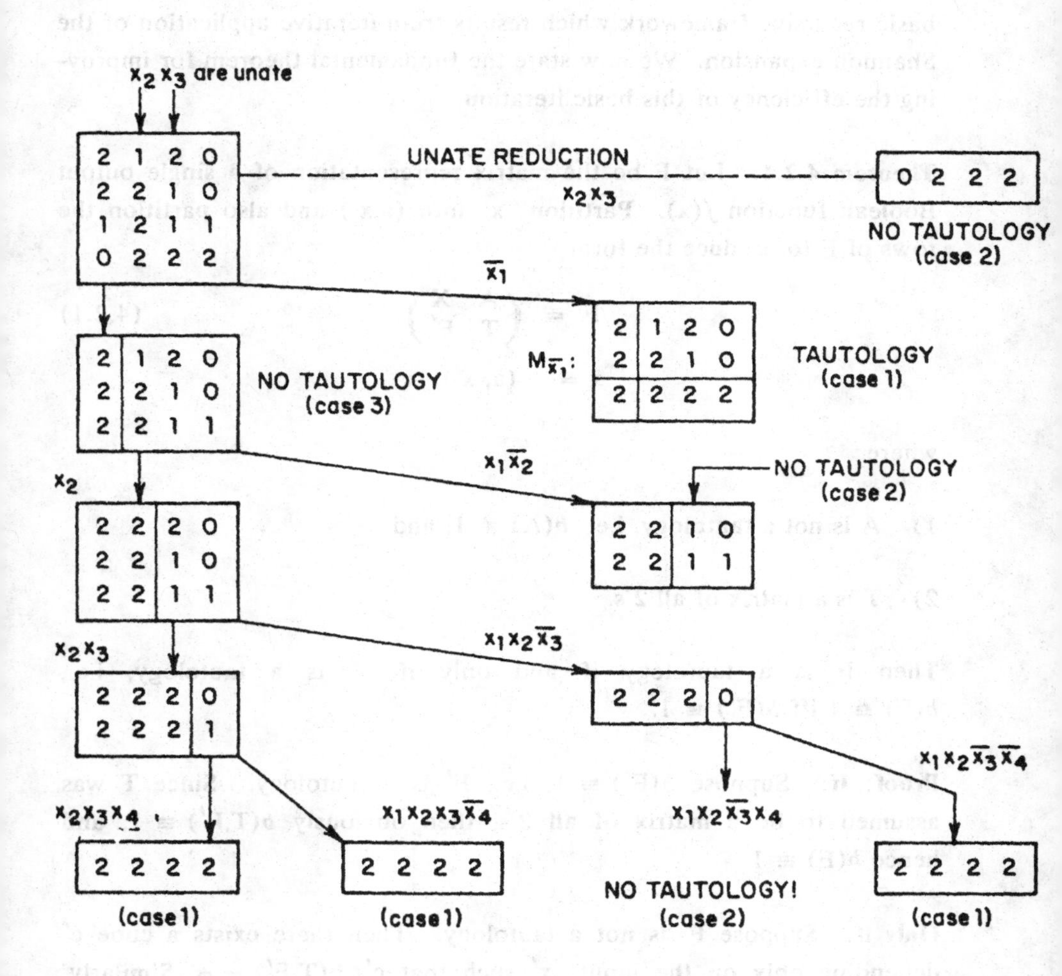

Figure 4.2.2

a row of all 2's, and the corresponding node is a leaf of the recursion tree.

4.2.2. Efficiency Results for Tautology.

The vanilla tautology algorithm of Section 4.2.1 establishes the basic recursive framework which results from iterative application of the Shannon expansion. We now state the fundamental theorem for improving the efficiency of this basic iteration.

Theorem 4.2.1: Let F be the matrix representation of a single output Boolean function $f(x)$. Partition x into (a,x′) and also partition the rows of F to produce the form

$$F = \begin{pmatrix} A & X \\ T & F' \end{pmatrix}, \qquad (4.2.1)$$

$$x = (a, x'),$$

where

1) A is not a tautology, i.e. $b(A) \not\equiv 1$, and

2) T is a matrix of all 2's.

Then F is a tautology if and only if F′ is a tautology, i.e., $b(F) \equiv 1$ iff $b(F') \equiv 1$.

Proof: **If**: Suppose $b(F') \equiv 1$, i.e., F′ is a tautology. Since T was assumed to be a matrix of all 2's, then obviously $b(T,F') \equiv 1$, and hence $b(F) \equiv 1$.

Only if: Suppose F′ is not a tautology. Then there exists a cube c′ depending only on the inputs x′ such that $c' \cap b(T,F') = \phi$. Similarly, since A is not a tautology there exists a cube c depending only on the inputs a such that $c \cap b(A,X) = \phi$. Then $c \cap c' \neq \phi$, but $c \cap c' \cap (b(A,X) + b(T,F')) = c \cap c' \cap b(F) = \phi$ and F is not a tautology.

■

A related result depends on the sparsity structure of the so-called personality matrix of F. Let P(F) be defined by

$$P_{ij} = \begin{cases} 1 & \text{if } F_{ij} \neq 2, \\ 0 & \text{otherwise.} \end{cases} \tag{4.2.2}$$

Suppose we discover row and column permutations R^T and C which produce the following block diagonal form.

$$R^T PC = \begin{matrix} P^1 & 0 & . & 0 \\ 0 & P^2 & . & \vdots \\ \vdots & \vdots & . & 0 \\ 0 & 0 & 0 & P^k \end{matrix} \tag{4.2.3}$$

where the P^i cannot be further decomposed in the same manner. Then we have the corresponding form,

$$R^T FC = \begin{matrix} F^1 & 2 & . & 2 \\ 2 & F^2 & . & \vdots \\ \vdots & \vdots & . & 2 \\ 2 & 2 & 2 & F^k \end{matrix} \tag{4.2.4}$$

for F, where "2" corresponds to a block matrix of all 2's.

Theorem 4.2.2: Suppose that a Boolean matrix F has the form (4.2.4). Then F is a tautology if and only if F^i is a tautology, for some $i \in \{1,2,..., k\}$.

Proof: The if part is trivial. To prove the only if part, assume that none of the F^i are tautologies; we will show F is not a tautology, by induction on k. This is trivial if $k=1$; assume it for $k-1$. Then (4.2.4) has the form (4.2.1) with $F' = F^k$ and A the matrix obtained by stripping away the last row and column blocks. By induction hypothesis, A is not a tautology; since F^k is also not a tautology, neither is F, by Theorem 4.2.1. ■

4.2.3. Improving the Efficiency of the Tautology Algorithm.

The results obtained in the previous section have proven to be useful in making the tautology algorithm more efficient. In the TAUTOLOGY algorithm, the effect of these results is packaged into subalgorithms UNATE__REDUCTION and COMPONENT__REDUCTION.

UNATE__REDUCTION applies Theorem 4.2.1 and the tautology test for a unate function (Proposition 3.3.4) as follows. If F has no unate columns, no action is taken. If there are unate columns, we declare F *not* a tautology if no row of 2's exists in these columns. If a row of 2's does exist, we determine F' as in Theorem 4.2.1, replace F by F' and recur until there are no unate columns. This has the effect of splitting *many* columns (i.e. all *unate* columns) in one operation, leading in some cases to dramatic improvements.

An example of this procedure is given below. The given matrix F is shown with horizontal and vertical lines indicating the partitioning of F into A^1, X^1, T^1, and F^1. Note A^1 is unate in this example. The procedure recurs once, and F is a tautology if and only if F^2 is a tautology (which it is).

$$F = \begin{array}{cc|cc} 1 & 2 & 0 & 1 \\ 2 & 1 & 1 & 0 \\ 1 & 1 & 1 & 2 \\ \hline 2 & 2 & 1 & 1 \\ 2 & 2 & 2 & 0 \\ 2 & 2 & 2 & 1 \end{array} = \begin{bmatrix} A^1 & X^1 \\ T^1 & F^1 \end{bmatrix}$$

$$F^1 = \begin{array}{c|c} 1 & 1 \\ \hline 2 & 0 \\ 2 & 1 \end{array} = \begin{bmatrix} A^2 & X^2 \\ T^2 & F^2 \end{bmatrix}$$

Similarly, the application of this method to the matrix M in Figure 4.2.2 allows one to quickly see that that example is not a tautology.

Theorem 4.2.2 is used in an analagous fashion by COMPONENT_REDUCTION. If the personality matrix of F is sufficiently sparse, we check for a nontrivial block structure. If it exists, F is decomposed into $F^1, F^2, \ldots, F^k$, and the F^i are tested individually for tautology, $i = 1,2,\ldots,k$. If any one of the F^i is a tautology, then by Theorem 4.2.2, F is a tautology. This can have the effect of drastically reducing the size of the space we must check for tautology.

4.2.4. Tautology for Multiple-Output Functions.

Suppose we are given a matrix pair (G,H) where $G = I(M(\mathcal{G}))$ and $H = O(M(\mathcal{G}))$. Let m be the number of outputs of

$g = b(M(\mathscr{G}))$. Define $R^j = \{i \mid H_{ij} = 4\}$ and $G^{[j]}$ be the submatrix of G formed by the rows R^j.

Remark: Note that according to the definition of tautology given in Chapter 2,

$$g \equiv b(G,H) \equiv 1 \text{ iff } g_j \equiv b(G^{[j]}) \equiv 1, \quad j = 1,\ldots, m. \tag{4.2.5}$$

That is, a multiple output Boolean function is a tautology if and only if each of its component Boolean functions is a tautology.

We illustrate these remarks with the following example. Suppose

$$(G,H) = \begin{matrix} 1 & 2 & 0 & 1 & & 3 & 4 \\ 2 & 1 & 1 & 0 & & 4 & 3 \\ 1 & 1 & 1 & 0, & & 4 & 4 \\ 2 & 2 & 2 & 0 & & 4 & 3 \\ 2 & 2 & 2 & 1 & & 4 & 4 \end{matrix} \tag{4.2.6}$$

Then

$$G^{[1]} = \begin{matrix} 2 & 1 & 1 & 0 \\ 1 & 1 & 1 & 0 \\ 2 & 2 & 2 & 0 \\ 2 & 2 & 2 & 1 \end{matrix}$$

$$G^{[2]} = \begin{matrix} 1 & 2 & 0 & 1 \\ 1 & 1 & 1 & 0 \\ 2 & 2 & 2 & 1 \end{matrix} \tag{4.2.7}$$

Theorem 4.2.1 can be applied to $G^{[1]}$ and $G^{[2]}$ above to show that $G^{[1]}$ is a tautology, $b(G^{[1]}) \equiv 1$, but $G^{[2]}$ is not a tautology, $b(G^{[2]}) \not\equiv 1$. Thus (G,H) is not a multiple-output tautology. We note, however, that Theorem 4.2.1 generalizes as follows to the multiple-output case, which permits us to do more than just reduce the multiple-output functions to a set of single-output functions and then check these functions, individually, for tautology.

Theorem 4.2.3: Let (G,H) be a matrix representation of a multiple-output logic function. Partition x into (a,x′) and partition the rows of $f(x) = b(G,H)$ to produce the form

$$(G,H) = \left(\begin{bmatrix} A & X \\ T & G' \end{bmatrix}, \begin{bmatrix} H^A \\ H' \end{bmatrix} \right), \qquad (4.2.8)$$

$$x = [a, x']$$

where

1) $b(A) \not\equiv 1$

2) T is matrix of all 2's.

Then $b(G,H) \equiv 1$ if and only if $b(G',H') \equiv 1$.

(The proof, similar to that of Theorem 4.2.1, is omitted.)

We can apply Theorem 4.2.3 to the example of (4.2.6) as follows. Partition G and H into

$$G = \begin{bmatrix} A & X \\ T & G' \end{bmatrix} = \begin{matrix} 1\ 2\ 0 & 1 \\ 2\ 1\ 1 & 0 \\ 1\ 1\ 1 & 0 \\ & \\ 2\ 2\ 2 & 0 \\ 2\ 2\ 2 & 1 \end{matrix},$$

$$(4.2.9)$$

$$H = \begin{bmatrix} H^A \\ H' \end{bmatrix} = \begin{matrix} 3\ 4 \\ 4\ 3 \\ 4\ 4 \\ \\ 4\ 3 \\ 4\ 4 \end{matrix}.$$

By Theorem 4.2.3, (G,H) is a tautology if and only if

$$(G',H') = \left(\begin{bmatrix} 0 \\ 1 \end{bmatrix}, \begin{bmatrix} 4 & 3 \\ 4 & 4 \end{bmatrix} \right), \qquad (4.2.10)$$

is a tautology, since T is a matrix of 2's, and

$$(A,H^A) = \begin{bmatrix} 1 & 2 & 0 \\ 2 & 1 & 1 \\ 1 & 2 & 1 \end{bmatrix}, \begin{bmatrix} 3 & 4 \\ 4 & 3 \\ 4 & 4 \end{bmatrix} \quad (4.2.11)$$

is *not* a tautology because A is not the single-output tautology. But (4.2.10) is a simpler problem than (4.2.6). In fact, the tautology question for (4.2.10) is solved by splitting once (as described below) and calling a modified version of SPECIAL_CASES.

Our implementation of the multiple-output tautology algorithm follows the same lines as that for the single-output case, so we will not give a multiple-output version of the algorithm of Figure 4.2.1. Instead, we describe below some of the main differences. SPECIAL_CASES is similar to the single-output case. For example, if $f = b(G,H)$, then f *is not* a tautology if some column of G has all 0's or all 1's, and *is* a tautology if every column of H has a 4 in a row which is all 2's in the corresponding row of G. If H has a column of all 3's, then f is (trivially) *not* a tautology. If G has a row of all 2's, the corresponding outputs with 4's as entries can be deleted. This may lead to rows with outputs of all 3's which can be deleted also. This may have the benefit of decreasing the number of rows and columns in the matrix G'.

As in the single-output case, we take special action if there are fewer than 7 inputs. In this event we break f into individual functions and answer the tautology question using special cases 1-4 of Figure 4.2.1. If any individual functions fail the tautology test, f is *not* a tautology.

Multiple-output UNATE_REDUCTION is quite similar to the single-output version. The only significant difference is that the output matrix, H, must be partitioned along with G. But again the approach is to apply Theorem 4.2.3 and the multiple-output version of Proposition 3.3.4 iteratively until a submatrix F^k of F is found that has no unate columns.

If there are more than 7 inputs and no unate variables we pass control to the multiple output version of COMPONENT_REDUCTION. If $f = b(G,H)$, we first find the components of G as in Theorem 4.2.2. For each component G^i, we then ask the tautology question separately for each of the single outputs

f_j. Now f_j is a tautology only if it is a tautology for some component G^i, and f is a tautology only if each f_j is, so we have answered the tautology question for f.

If SPECIAL__CASES, UNATE__REDUCTION, and COMPONENT__REDUCTION all return $T = -1$, then, as in the single-output case, a splitting variable is selected. The multiple-output version of algorithm TAUTOLOGY is called recursively, and the whole process described above starts over again, but this time with smaller matrices (G^k, H^k) (cf. Figure 4.2.2).

4.3 Expand

In this section, we describe the procedure $\mathscr{F} \leftarrow$ EXPAND $(\mathscr{F}, \mathscr{R})$ (Figure 4.3.1) which transforms a given cover $\mathscr{F}$ of ff into a prime cover.

The expansion process, carried out one cube at a time by EXPAND1 (Figure 4.3.2), is guided by the goal of removing as many cubes as possible from $\mathscr{F}$. This goal is accomplished heuristically by processing the cubes of $\mathscr{F}$ sequentially and by maximizing the number of cubes covered at each step of the expansion process. As we replace each cube $\mathscr{F}^i$ in $\mathscr{F}$ by a prime cube $(\mathscr{F}^i)^+$, containing it, we delete those cubes of $\mathscr{F}$ contained in $(\mathscr{F}^i)^+$. As a result, the final cover is also minimal with respect to single cube containment.

The result of EXPAND depends on the order in which the cubes are expanded. The cubes which are most likely to cover other cubes and not be covered by other cubes should be expanded first. MINI[HON 74] tries to identify "essential cubes," i.e. cubes covering vertices not covered by other cubes in $\mathscr{F}$, and orders them first. We arrange the cubes of $\mathscr{F}$ in order of decreasing size. The rationale of our choice is simply that large cubes are more likely to cover other cubes and less likely to be covered by other cubes. This ordering is very simple, but in most examples it is at least as effective as the more complex MINI strategy.

The heart of EXPAND is the procedure $(W, c^+) \leftarrow$ EXPAND1$(c, \mathscr{F}, \mathscr{R})$ which expands a cube $c = \mathscr{F}^i$ of a cover $\mathscr{F} = \{\mathscr{F}^1, \mathscr{F}^2, \ldots\}$, into a prime implicant, c^+ of ff and lists in W the cubes in $\mathscr{F}$ covered by c^+. The prime $c^+ \supseteq c$ is chosen so that it covers

```
Procedure EXPAND (𝓕, 𝓡)

/* Given 𝓕 a cover of ff = (f, d,r), and
/* 𝓡, a cover of r,
/* returns a prime cover of ff.

    Begin
    𝓕 ← DECREASING_ORDER (𝓕)          /* The cubes in 𝓕 are reordered
                                        /* in decreasing size, i.e. largest
                                        /* cubes are processed first.
    for (j = 1 ,..., |𝓕|)
        Begin
        (W, 𝓕^j) ← EXPAND1 (𝓕^j, 𝓡, 𝓕)
        𝓕 ← (𝓕 ∪ {𝓕^j}) − W
        End
    return (𝓕)
    End
```

Figure 4.3.1

as many cubes as possible of $\mathcal{F}$. We use the following notation.

$\mathcal{F}$: a cover of ff (it covers the onset of ff and it may cover some vertices of the don't care set of ff)

$\mathcal{R}$: a cover of the off − set of ff (it covers only the off − set of ff)

$F = M(\mathcal{F})$: a $0-1-2-3-4$ matrix representation of $\mathcal{F}$ as defined in Chapter 2.

Given a cube $c = \mathcal{F}^i$ to be expanded, an "optimally expanded" prime cube $d \supseteq c$ is defined as a cube with the property

$$|\{j \mid \mathcal{F}^j \subseteq d\}| \text{ is maximum and,} \tag{4.3.1a}$$

$$\text{the number of literals of d is minimum.} \tag{4.3.1b}$$

Property (4.3.1a) expresses our **primary** objective of minimizing the number of cubes in the expanded cover. Property (4.3.1b), expresses our secondary objective; that the prime cube d be as large as possible

```
Procedure EXPAND1(c, 𝓕, 𝓡)

/* Given 𝓕 a cover of ff = (f,d,r)
/* 𝓡 a cover of r
/* and c ≡ 𝓕^i, a cube of 𝓕,
/* returns c+ ⊇ c, an optimal-coverage prime implicant and
/* W, the set of cubes of 𝓕 contained in c+.

Begin
(𝔹,ℂ) ← MATRICES (c,𝓕,𝓡)                      /* Construct blocking matrix 𝔹
                                              /* and covering matrix ℂ

𝕃 ← φ; ℝ ← φ                                  /* Initialize Lowering set 𝕃
                                              /* and Raising set ℝ.

N ← NCOLS(𝔹)                                  /* N equals the initial number
                                              /* of columns of 𝔹

while ( |𝕃| + |ℝ| < N AND 𝔹 ≠ φ AND ℂ ≠ φ)
      Begin
      X_E ← ESSENTIAL(𝔹)                      /*Append essential columns to 𝕃
      𝕃 ← 𝕃 ∪ X_E
      (𝔹,ℂ) ← ELIM1 (𝔹, ℂ, X_E)               /* Eliminate columns X_E
                                              /* and associated rows in 𝔹,ℂ

      J^i* ← MFC(𝔹,ℂ)                         /* Compute Maximum Feasible
                                              /* Covering set.
      if ( |J^i*| = 0) J^i* ← EG(ℂ)           /* If none ∃ choose
                                              /* J^i* according to End Game

      X_I ← INESSENTIAL(𝔹)                    /* Augment raising set. Inessential
      ℝ ← ℝ ∪ J^i* ∪ X_I                      /* columns are zero columns of 𝔹

      (𝔹,ℂ) ← ELIM2 (𝔹, ℂ, J^i* ∪ X_I)        /* Eliminate columns J^i* ∪ X_I
                                              /* and zero rows of ℂ
      End
if (𝔹 ≠ φ) then 𝕃 ← MINLOW(𝔹)                 /* Third case of end game
W ← {𝓕^i ∈ 𝓕 | ℂ_ij = 0 for all j ∈ 𝕃}        /* W is the set of cubes
return (W, c+(𝕃, c))                          /* contained in c+
End
```

Figure 4.3.2

i.e., depend on the least number of variables. The expand operation can be regarded as a "greedy" optimization step which attempts a "local" minimization of a weighted sum of the number of literals in $\mathcal{F}$ and the number of cubes in $\mathcal{F}$. Of course if we were willing to enumerate all the prime implicants $p \supseteq c$ of F, we could select one, d, which best satisfies (4.3.1). Since this is usually prohibitively expensive, we proceed heuristically, and obtain a prime expanded cube, $c^+ \geq c$, which approximately satisfies (4.3.1). The strategy is guided by a "blocking matrix", $\mathbb{B}$, and a "covering matrix", $\mathbb{C}$.

4.3.1. The Blocking Matrix

For clarity in this and the next section we will confine our discussions to the single-output case, and in Section 4.3.3 extend the concepts to the multiple-output case.

The blocking matrix, $\mathbb{B}$, is a 0-1 matrix determined by the cube $c = \mathscr{F}^k$ to be expanded and by $\mathscr{R}$, the cover of the off-set. The rows of $\mathbb{B}$ are in one to one correspondence with the cubes $\mathscr{R}^i$. The elements of $\mathbb{B}$ are defined by

$$\mathbb{B}_{ij} = \begin{cases} 1 & \text{if } \{(c_j = 1) \text{ and } (M(\mathscr{R})_{ij} = 0) \text{ or} \\ & \quad (c_j = 0) \text{ and } (M(\mathscr{R}_{ij}) = 1\} \\ 0 & \text{otherwise.} \end{cases} \qquad (4.3.2a)$$

(An "a" version of a formula in these chapter will refer to the single-output case; in Section 4.3.3 the corresponding multiple-output formula will be labelled "b".) The notation $c_j = 1$ ($c_j = 0$) means that the cube to be expanded contains the literal x_j ($\bar{x}_j$).

Our purpose in the EXPAND operation is to select an optimal set of variables (corresponding to columns of $\mathbb{B}$), called the "lowering set" $\mathbb{L}$. The lowering set $\mathbb{L}$ defines the expanded cube c^+ by

$$c^+(\mathbb{L}, c)_j = \begin{cases} c_j, & j \in \mathbb{L}, \\ 2 & \text{otherwise} \end{cases} \qquad (4.3.3a)$$

$\mathbb{L}$ is called a lowering set because procedurally we regard each variable "present" in c (i.e. not 2 in the matrix representation of $\mathscr{F}$) as being potentially raised to "2" in the expanded cube, c^+. Variables in $\mathbb{L}$ are then "lowered" to their original values in c.

It is clear that $c^+ \geq c$; furthermore, $c^+(\mathbb{L}, c)$ is an implicant of $\mathscr{f\!f}$ if and only if $\mathbb{L}$ is a column covering of $\mathbb{B}$ (Here "column covering" means that every row of $\mathbb{B}$ contains a 1 in some column which appears in $\mathbb{L}$). Indeed, $\mathbb{L}$ covers the ith row of $\mathbb{B}$ if and only if there is some variable j such that $c^+(\mathbb{L}, c)_j = 0$ and $\mathscr{R}^i_j = 1$, or vice-versa. Equivalently, $\mathbb{L}$ covers the ith row of $\mathbb{B}$ exactly when c^+ does not meet the ith cube of the off-set $\mathscr{R}$ of ff. If $\mathbb{L}$ is a covering of $\mathbb{B}$, then c^+ does not meet the off-set at all, and hence c^+ is an implicant of ff.

Thus coverings $\mathbb{L}$ of $\mathbb{B}$ are in one-to-one correspondence with implicants $c^+ \supseteq c$. Furthermore, $\mathbb{L}$ is a minimal cover of $\mathbb{B}$ if and only if the corresponding c^+ is a prime implicant.

For example, suppose $\mathscr{F} = \{x_1\bar{x}_2x_3, \bar{x}_1\bar{x}_3, \bar{x}_1\bar{x}_2, x_1\bar{x}_2\bar{x}_3\}$, $c = \bar{x}_1\bar{x}_2 \equiv \mathscr{F}^3$, and $\mathscr{R} = \{x_1x_2, x_2x_3\}$. In matrix form we have

variables:	x_1	x_2	x_3
c:	0	0	2
$M(\mathscr{R})$:	1	1	2
	2	1	1
$\mathbb{B}(\mathscr{R},c)$:	1	1	0
	0	1	0

(4.3.4)

In this example, the unique minimal lowering set is {2}, since column 2 has a "1" in each row of $\mathbb{B}$. The corresponding expanded cube is $c^+ = \bar{x}_2$, or 202 in matrix form. Because variable x_2 is lowered (i.e., retains its original value of "0"), c^+ is orthogonal to the two terms of $\mathscr{R}$.

Since the number of literals in the cube c^+ is determined by the cardinality of $\mathbb{L}$, the following observation is immediate.

Proposition 4.3.1. (Maximum Prime): If $\mathbb{L}$ is a minimum column-cover of $\mathbb{B}$, then $c^+(\mathbb{L}, c)$ is a largest prime implicant of ff containing c. ($\mathbb{L}$ and c^+ need not be unique) . ■

The problem of finding the optimal choice for c^+ can now be converted into the problem of finding a covering $\mathbb{L}$ for $\mathbb{B}$ which satisfies some additional criterion. For example, by the above proposition, to obtain a c^+ with a minimum number of literals, we must construct an $\mathbb{L}$ of minimum cardinality. Such a c^+ would optimize our secondary objective, but not necessarily our primary one, to cover other cubes of $\mathscr{F}$. So instead we focus on the latter property. Our search for $\mathbb{L}$ is guided by another auxiliary construct, the covering matrix, $\mathbb{C}$.

4.3.2 The Covering Matrix

The covering matrix, $\mathbb{C}$, is determined by the given cover $\mathcal{F}$ and the cube to be expanded, c. The covering matrix is used to guide the selection of a minimal lowering set $\mathbb{L}$ such that the corresponding c^+ covers as many cubes in $\mathcal{F}$ as possible. To this end, we define $\mathbb{C}$ as follows:

$$\mathbb{C}_{ij} = \begin{cases} 1 & \text{if } \{(c_j = 1) \text{ AND } (M(\mathcal{F})_{ij} \neq 1)\} \text{ OR} \\ & \{(c_j = 0) \text{ AND } (M(\mathcal{F})_{ij} \neq 0)\}, \\ 0 & \text{otherwise.} \end{cases} \tag{4.3.5a}$$

If $\mathbb{L}$ is a lowering set, then the kth cube of the cover $\mathcal{F}^k$ is covered by $c^+(\mathbb{L},c)$ if and only if $\mathbb{C}_{kj} = 0$, $\forall\ j \in \mathbb{L}$. So our objective is to select a minimal set of columns $\mathbb{L}$ covering every row of $\mathbb{B}$ but as *few* rows of $\mathbb{C}$ as possible. As usual, row i of $\mathbb{C}$ is covered by $\mathbb{L}$ if $\mathbb{C}_{i\ell} = 1$ for some $\ell \in \mathbb{L}$.

In example (4.3.4), we have

$$\begin{array}{rccc}
\text{variables:} & x_1 & x_2 & x_3 \\
\\
c: & 0 & 0 & 2 \\
\\
M(\mathcal{F}): & 1 & 0 & 1 \\
 & 0 & 2 & 0 \\
 & 0 & 0 & 2 \\
 & 1 & 0 & 0 \\
\\
\mathbb{C}(\mathcal{F},c): & 1 & 0 & 0 \\
 & 0 & 1 & 0 \\
 & 0 & 0 & 0 \\
 & 1 & 0 & 0
\end{array} \tag{4.3.6}$$

The lowering set $\mathbb{L} = \{2\}$ consists of a single column which has 1's only in the second row of $\mathbb{C}$. Thus $\mathcal{F}^1$, $\mathcal{F}^3$ and $\mathcal{F}^4$ are covered by the expanded cube $c^+ = 202 = \bar{x}_2$. (Of course if c is $\mathcal{F}^i$, then c^* always covers $\mathcal{F}^i$; in this example i = 3.)

We shall describe below the procedure for using $\mathbb{C}$ to guide our selection of $\mathbb{L}$. But first we describe how the construction of $\mathbb{B}$ and $\mathbb{C}$ is to be done in the multiple output case.

4.3.3 Multiple-Output Functions

Our approach to the multiple-output case is to represent $\mathscr{R}$ in "unwrapped" form. That is, each term in $\mathscr{R}$ contributes to only one output. Any multiple output implicant affecting $k>1$ outputs, must be replaced by k replicas, each with the same input part but each with only a single "4" in the output part. For example, the multiple-output implicant 21203434 must be split into two single output implicants 21203433 and 21203334. This unravelled representation of $\mathscr{R}$ makes the computation of the multiple-output blocking matrix very simple.

We augment the definition (4.3.2a) of $\mathbb{B}$ as follows, assuming j represents an output component, by defining

$$\mathbb{B}_{ij} = \begin{cases} 1 & \text{if } \{(c_j = 3) \text{ AND } (M(\mathscr{R})_{ij} = 4)\}, \\ 0 & \text{otherwise.} \end{cases} \tag{4.3.2b}$$

Note that the cube to be expanded, $c \in \mathscr{F}$, need not be unwrapped. It suffices to represent $\mathscr{R}$ without term sharing, and use (4.3.2b) to construct $\mathbb{B}$

We similarly extend the definition of $c^+(\mathbb{L},c)$ to apply to output columns. Assuming j represents an output column, we define

$$c^+(\mathbb{L}, c)_j = \begin{cases} c_j, & j \in \mathbb{L} \\ 4 & \text{otherwise.} \end{cases} \tag{4.3.3b}$$

Proposition 4.3.2: If $\mathbb{B}_{ij} = 1$, then

$$c^+(\{j\}, c) \cap \mathscr{R}^i = \phi.$$

Proof: Note that $c^+(\{j\},c)_j = c_j$. If j represents an input column, then $\mathbb{B}_{ij} = 1$ only if c and $\mathscr{R}^i$ conflict in the jth variable i.e., $c_j = 1$, $M(\mathscr{R})_{ij} = 0$, or $c_j = 0$, $M(\mathscr{R})_{ij} = 1$. Since $c_j^+ = c_j$, c^+ and $\mathscr{R}^i$ are disjoint. On the other hand, if j represents an output column, then $\mathbb{B}_{ij} = 1$ only if $c_j = c_j^+ = 3$ and $M(\mathscr{R})_{ij} = 4$. Since $\mathscr{R}$ is "unwrapped", $M(\mathscr{R})_{ik} = 3$, $\forall\ k \neq j$. Thus the outputs of c^+ and $\mathscr{R}^i$ are disjoint, and $c^+ \cap \mathscr{R}^i = \phi$. ∎

Thus a "1" in the blocking matrix $\mathbb{B}$ has the same meaning (a reason for orthogonality) regardless of whether j is an input column or an output column. This would not be true if the representation of $\mathscr{R}$ allowed term sharing.

If there is substantial potential term sharing in the representation of $\mathscr{R}$, the unwrapped representation causes increased storage requirements. However, this disadvantage is offset by the simplicity and efficiency of the expansion algorithm, which can now treat input and output columns in an identical manner. Because there is no asymmetry between inputs and outputs, the expansion algorithm also produces better results.

The covering matrix, $\mathbb{C}$ is similarly modified in the multiple-output case, except in this case there is no need to "unwrap" the matrix representation of the original cover, $\mathscr{F}$. Instead, we simply extend the definition of $\mathbb{C}_{ij}$ as follows: when j corresponds to an output part component,

$$\mathbb{C}_{ij} = \begin{cases} 1 & \text{if } \{(c_j = 3) \text{ AND } (M(\mathscr{F})_{ij} = 4)\} \\ 0 & \text{otherwise,} \end{cases} \tag{4.3.6b}$$

Proposition 4.3.3: If $\mathbb{C}_{ij} = 1$, then

$$c^+(\{j\}, c) \not\subseteq \mathscr{F}^i.$$

Proof: As before, $c_j^+ = c_j$. If j corresponds to an input column, then $\mathbb{C}_{ij} = 1$ only if $c_j = c_j^+ = 1$ and $M(\mathscr{F})_{ij} \neq 1$ or $c_j = 0$ and $M(\mathscr{F})_{ij} \neq 0$. In either case, $c^+ \not\subseteq \mathscr{F}^i$. If j corresponds to an output column, then $\mathbb{C}_{ij} = 1$ only if $c_j = c_j^+ = 3$ and $M(\mathscr{F})_{ij} = 4$. Again $c^+ \not\subseteq \mathscr{F}^i$. ■

Roughly speaking, a "1" in $\mathbb{C}_{ij}$, means "a reason for noncovering", regardless of whether j is an input or output column.

To illustrate the above definitions, consider the following example. Suppose

$$M(\mathscr{F}) = \begin{matrix} 22143 \\ 12244, \\ 21134 \end{matrix} \quad M(\mathscr{R}) = \begin{matrix} 02043 \\ 02034 \\ 00234 \end{matrix} \tag{4.3.8}$$

where $\mathscr{R}$ is represented with no term sharing. Using (4.3.2) and

(4.3.5), respectively, with $c = \mathcal{F}^3 = 21134$, we obtain

$$\mathbb{C} = \begin{matrix} 01010 \\ 01110, \\ 00000 \end{matrix} \quad \mathbb{B} = \begin{matrix} 00110 \\ 00100. \\ 01000 \end{matrix} \tag{4.3.9}$$

Note that in this example the only possible choice for $\mathbb{L}$ is {2,3}, i.e., the second and third columns constitute the only lowering set. Thus $c^+ = 21144$. Also, since the second column of $\mathbb{C}$ has a 1 in both the first and second rows, we conclude that the "expanded" cube c^+ covers only itself and none of the other cubes in $\mathcal{F}$. Notice also that column 4 was not lowered implying that c was not prime. This illustrates how an output can be "raised" during the expansion to prime.

4.3.4 Reduction of the Blocking and Covering Matrices

In the procedure EXPAND1, the lowering set $\mathbb{L}$ is built up by adding columns to it one by one. Each time a new column is added, the blocking and covering matrices can be reduced in size to ensure efficient processing. Once column j has been selected for inclusion in $\mathbb{L}$, this column may be deleted from $\mathbb{B}$ and $\mathbb{C}$. Then all rows of $\mathbb{B}$ and $\mathbb{C}$ with a 1 in this column may also be deleted. For $\mathbb{B}$, this means that disjointness from the corresponding rows of the off-set has now been ensured. For $\mathbb{C}$, this means that it is no longer possible for the partially expanded cube to ever cover (i.e., contain) any of the corresponding cubes in $\mathcal{F}$. We refer to this deletion process as the "elimination" of rows and columns and will denote it in the EXPAND1 procedure by ELIM1 (cf., Figure 4.3.2).

Note that some columns must be included in $\mathbb{L}$ to ensure the disjointness of c^+ and $\mathcal{R}$. In fact, if a row of $\mathbb{B}$ has only one nonzero entry, say $\mathbb{B}_{ij}$, then j must appear in $\mathbb{L}$ since this is the only way this row can be covered. We denote the set of all such "Essential Columns" by X_E. That is, for the original $\mathbb{B}$

$$X_E = \{j \mid \mathbb{B}_{ij} = 1, \ \sum_k \mathbb{B}_{ik} = 1, \ \text{for some } i\}. \tag{4.3.10}$$

Note that $c^+(X_E, c)$ is the so-called "overexpanded cube" c* [HON 74], defined as the smallest cube which contains every prime implicant containing c. Note that c* need not itself be an implicant of $\mathscr{f}$, because some extra columns <u>not</u> in X_E may be required to ensure the disjointness of c^+ and $\mathscr{R}$. However, there is always freedom in the selection of these additional columns, while the set X_E is not determined by choice. In the example (4.3.9), $c^* = c^+$ since the set X_E constitutes a complete covering of $\mathbb{B}$, and therefore there is no need to include any additional columns in $\mathbb{L}$. The essential columns of $\mathbb{B}$ are determined by ESSENTIAL in the EXPAND1 procedure (cf. Figure 4.3.2).

4.3.5 The Raising Set and Maximal Feasible Covering Set

Once the essential columns of $\mathbb{B}$ have been chosen and the rows they cover have been eliminated, all remaining rows of $\mathbb{B}$ contain two or more nonzero entries. Hence we can always choose at least one column to "permanently raise", i.e. exclude from $\mathbb{L}$. We call the set of such permanently raised columns the **raising set** $\mathbb{R}$. Of course we cannot choose $\mathbb{R}$ so large that it covers **all** the 1's in any given row of $\mathbb{B}$, since every row must eventually be covered by **some** column in $\mathbb{L}$. In procedure EXPAND1, both $\mathbb{L}$ and $\mathbb{R}$ increase in cardinality until all of the columns are included in one or the other.

Whenever a set of columns is added to $\mathbb{R}$, we can eliminate these columns from $\mathbb{B}$ and $\mathbb{C}$. This may cause rows of all zeros in $\mathbb{C}$ which means the corresponding $\mathscr{F}^i$ are covered. In this case we eliminate these rows. Again we call this process "elimination" and refer to it in the EXPAND1 procedure as ELIM2.

Also at any stage we may have columns in $\mathbb{B}$ consisting of all zeros. These are inessential columns and may be included in $\mathbb{R}$. Zero columns come about initially because the columns of c which are already 2 or 4 will lead to inessential columns. They may also occur after a column is eliminated if the eliminated column dominates another one. The zero columns of $\mathbb{B}$ are found by subprocedure INESSENTIAL in EXPAND1.

The choice of $\mathbb{R}$ is guided by the desire to cover as many cubes $\mathscr{F}^i$ as possible. Note that if the current $\mathbb{B}$ has no row singletons, then any cube $\mathscr{F}^i$ corresponding to a row singleton of $\mathbb{C}$ may be covered by including in $\mathbb{R}$ the number of the column with a 1 in this row. This notion generalizes as follows:

Given $\mathbb{L}$ and $\mathbb{R}$, let J^i be the set of columns, not in $\mathbb{R}$ or $\mathbb{L}$, which must be included in $\mathbb{R}$ in order to cover cube $\mathscr{F}^i$. Let $\bar{J}^i$ be the remaining columns not in $\mathbb{L}$, $\mathbb{R}$, or J^i.

Definition: The cube $\mathscr{F}^i$ is **feasibly covered** (the set J^i is feasible) if for all k,

$$\left(\sum_{j \in \mathbb{L} \cup \bar{J}^i} \mathbb{B}_{kj} \geq 1 \right) \tag{4.3.11}$$

This means that the set J^i can be permanently raised (included in $\mathbb{R}$) and there will remain enough columns in $\bar{J}^i$ to complete $\mathbb{L}$ to a cover of $\mathbb{B}$

Our strategy is to select one of the feasibly covered cubes $\mathscr{F}^{i^*}$ which is maximal in the following sense:

$$i^* \in \underset{J^i \text{ feasible}}{\operatorname{argmax}} \; |\{k \mid J^i \supseteq J^k\}| . \tag{4.3.12}$$

Ties are broken by choosing the minimum $|J^{i^*}|$. Note that since J^i is feasible, any $J^k \subseteq J^i$ is also feasible. Thus i* is chosen so that as many cubes $\mathscr{F}^i$ as possible are covered, and at the same time a minimal sufficient set of columns is committed to the permanently raised set $\mathbb{R}$, thus keeping free subsequent choices for possibly covering other cubes later on. Ideally one should consider all feasible sets, not just those J^i associated with cubes $\mathscr{F}^i$, and choose the smallest one which contains the most J^i. However, we found that in the typical case only a few cubes are feasibly covered, and the additional complication would be typically without benefit.

In the EXPAND1 procedure shown in Figure 4.3.2, if there exist any feasibly covered cubes, we select i* and add J^{i^*} to $\mathbb{R}$;

$$\mathbb{R} \leftarrow \mathbb{R} \cup J^{i^*}.$$

The corresponding rows and columns of $\mathbb{B}$ and $\mathbb{C}$ are then "eliminated". This may lead to more "essential columns". We therefore include these in $\mathbb{L}$, eliminate the corresponding columns and associated rows from $\mathbb{B}$ and $\mathbb{C}$ and choose another feasible cube. These procedures are iterated until there are no more feasibly covered cubes or until all columns have been selected for either $\mathbb{L}$ or $\mathbb{R}$. In the later case, we are done. If there are no more feasibly covered cubes, but some columns have not been selected, we enter the "Endgame".

4.3.6 The Endgame

In the Endgame, no more cubes in $\mathcal{F}$ may be covered in the expansion of c, but some columns remain to be chosen. There are three cases:

1. $\mathbb{B} \neq \phi$, $\mathbb{C} \neq \phi$: Here no further cubes $\mathcal{F}^i$ can be covered by c^+, no matter how we complete the selection of $\mathbb{L}$. We attempt to "spend" the remaining degrees of freedom using subprocedure EG($\mathbb{C}$). In EG, we try to increase the likelihood that cubes partially covered by the expanded cube we are now processing can be cooperatively covered by this and subsequently expanded cubes, and therefore eliminated by a subsequent IRREDUNDANT_COVER step. EG chooses a single column which has maximum column count in $\mathbb{C}$. Note that at this point $\mathbb{B}$ has no row singletons so <u>any</u> single column of $\mathbb{C}$ can be added to $\mathbb{R}$. The motivation for this heuristic is that if $\mathbb{R} \leftarrow \mathbb{R} \cup EG(\mathbb{C})$, then all cubes $\mathcal{F}^k$ whose corresponding row $\mathbb{C}_k$ has a 1 in column EG($\mathbb{C}$) are "partially" covered by c^+. That is, if we cannot cover $\mathcal{F}^k$ completely we try to increase the overlap between c^+ and as many cubes of $\mathcal{F}$ as possible, hoping that the remainder of $\mathcal{F}^k$ will be covered by the expansion $\hat{c}^+$ of a subsequent cube $\hat{c}$.

2. $\mathbb{B} \neq \phi$, $\mathbb{C} = \phi$: In this case, our strategy is to choose $\mathbb{L}$ to minimize the number of literals in c^+, i.e. to minimize the size of $\mathbb{L}$. We call a minimum covering algorithm MINLOW (which is very similar to MINUCOV described in Section 4.5) to find a minimal lowering set $\mathbb{L}'$ with the property

$$\left(\sum_{j \in \mathbb{L}'} \mathbb{B}_{ij} \right) \geq 1, \ \forall i. \qquad (4.3.13)$$

Then we set $\mathbb{L} = \mathbb{L} \cup \mathbb{L}'$, $c^+ = c^+(\mathbb{L}, c)$ and terminate.

3. $\mathbb{B} = \phi$: In this case, there is no reason to lower any of the remaining columns so we set $c^+ = c^+(\mathbb{L}, c)$ and terminate.

4.3.7 The primality of c^+

We will show below that the cube c^+ produced by EXPAND1 is a prime implicant of $\mathcal{ff}$. Therefore it can never be covered by a subsequent expansion of a cube $\hat{c} \in \mathcal{F}$, and hence when expanding $\hat{c}$, we do not include a row for c^+ in our covering matrix $\mathbb{C}$

Theorem 4.3.4: The set $\mathbb{L}$ constructed in procedure EXPAND1 is a minimal lowering set, i.e., $c^+(\mathbb{L}, c)$ is a prime implicant of $\mathcal{ff}$.

Proof: There are only two ways that an index ℓ can be a member of $\mathbb{L}$ in procedure EXPAND1; it was either added as an **essential column** (for some partially eliminated $\mathbb{B}$), or it was added by MINLOW in the final step. If ℓ was added as an essential column, then, at the stage it was added, it covered a row singleton; therefore that row is covered by no subsequently chosen column, and hence ℓ cannot be eliminated from $\mathbb{L}$.

If ℓ was added by MINLOW, it is part of a minimal cover of the rows of $\mathbb{B}$ that had not yet been covered; since no further elements were added to $\mathbb{L}$ thereafter, again ℓ cannot be eliminated from $\mathbb{L}$. So in any case, no proper subset of $\mathbb{L}$ covers the rows of $\mathbb{B}$, and therefore $\mathbb{L}$ is minimal and c^+ is prime. ∎

The following example illustrates how the covering matrix $\mathbb{C}$ guides the expansion process in Procedure EXPAND1 and how output columns are treated on an equivalent basis with the inputs. Let

$$M(\mathcal{F}) = \begin{matrix} 0 & 1 & 0 & 4 & 3 \\ 1 & 0 & 1 & 4 & 4 \\ 0 & 1 & 0 & 3 & 4 \\ 0 & 1 & 1 & 4 & 3 \\ 1 & 2 & 0 & 4 & 4 \\ 0 & 0 & 1 & 3 & 4 \end{matrix}$$

where each of the six cubes (solid dots) is shown in Figure 4.3.3. Let $M(\mathcal{R})$, a matrix representation of the cover of the off-set, be

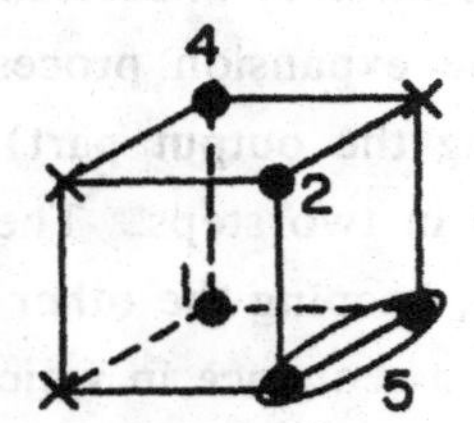

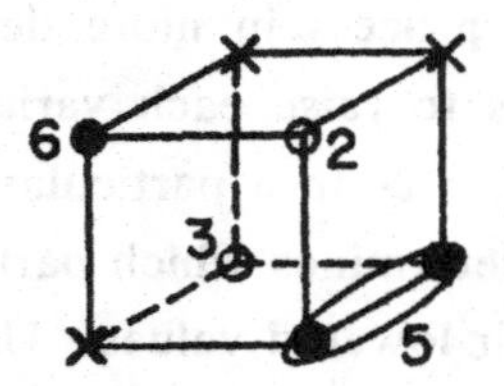

Figure 4.3.3 Cubes for Blocking and Covering Matrix Example

$$M(\mathcal{R}) = \begin{matrix} 1 & 1 & 1 & 4 & 3 \\ 2 & 1 & 1 & 3 & 4 \\ 0 & 0 & 0 & 3 & 4 \\ 0 & 0 & 2 & 4 & 3 \end{matrix} ,$$

also illustrated (with crosses) in Figure 4.3.3. Suppose we expand $c^1 = 0\ 1\ 0\ 4\ 3$. The blocking and covering matrices are

$$\mathbb{B} = \begin{matrix} 1 & 0 & 1 & 0 & 0 \\ 0 & 0 & 1 & 0 & 1 \\ 0 & 1 & 0 & 0 & 1 \\ 0 & 1 & 0 & 0 & 0 \end{matrix} , \quad \mathbb{C} = \begin{matrix} 0 & 0 & 0 & 0 & 0 \\ 1 & 1 & 1 & 0 & 1 \\ 0 & 0 & 0 & 0 & 1 \\ 0 & 0 & 1 & 0 & 0 \\ 1 & 1 & 0 & 0 & 1 \\ 0 & 1 & 1 & 0 & 1 \end{matrix} .$$

In expanding $c^1 = 0\ 1\ 0\ 4\ 3$, we first set $\mathbb{L} \leftarrow \{2\}$ since row 4 of $\mathbb{B}$ is a row singleton in column 2. At this point either c^3 or c^4 can be feasibly covered but, since column 5 has more 1's in $\mathbb{C}$ we set $\mathbb{R} \leftarrow \{5\}$. This creates, after ELIM2, a row singleton in row 2 of $\mathbb{B}$ forcing 3 into $\mathbb{L}$; $\mathbb{L} \leftarrow \{2,3\}$. At this point all of $\mathbb{B}$ is covered by $\mathbb{L}$. Hence $c^+(\{2,3\}) = 2\ 1\ 0\ 4\ 4$. Note that both column 1 and column 5 (i.e., input 1 and output 2), were "raised", illustrating how inputs and outputs are treated equivalently.

The expand algorithm which we have described in this section is conceptually new and represents a significant advance over the MINI cube expansion process. To appreciate this, we have to understand the MINI process in more detail. The MINI cube expansion process attempts to raise each variable or part (including the output part) of a cube $c \in \mathscr{F}$ in a particular order. This is done in two steps. The first step determines which parts can be raised alone, keeping the other parts at their lowered values. Using this information, a sequence in which the part raising will be attempted is determined, heuristically guided by the objective of covering as many of the other cubes of the cover as possible. Of course, not all parts in this sequence can be raised; only the first part will be guaranteed. The raising sequence is determined by forming the overexpanded cube c^*, which is obtained by raising each part that can be raised alone. This cube may not be in a cover of $\mathscr{ff}$, but it is used to find the set $\mathscr{Y}$ of cubes $\mathscr{F}^i$ of $\mathscr{F}$ which are candidates for being covered. Among the cubes $\mathscr{F}^i$ which are covered by c^*, MINI gives the kth part a weight equal to the number of cubes of $\mathscr{Y}$ whose k^{th} part is covered by c_k^*. The order of part raising is then done in decreasing order of the weights. This assignment of the sequence of part raising is statically determined before any parts are raised. In fact, after one part is raised, a repetition of the sequence determination mechanism may give rise to a different sequence. This is not done in MINI since it would be too expensive.

Our EXPAND process, on the other hand, is dynamic. The blocking and covering matrices are built initially, and these are used to precisely determine the consequences of raising a particular part. In addition, instead of choosing only one part at a time to be raised, we select according to equation (4.3.12) an optimal set J^{i^*} to be raised simultaneously, thus exercising a more global strategy. As sets of columns are chosen to be raised or lowered, the blocking and covering matrices are continually updated, so the effect of subsequent selections can be predicted exactly.

4.4 Essential Primes

Essential primes must be in all prime covers of $\mathcal{f}$. For this reason it is computationally desirable to eliminate essential primes from consideration when the operations EXPAND, REDUCE, and IRREDUNDANT_COVER are performed iteratively.

Recall that a prime p is **essential** if there is a minterm x in the on-set f which is contained in this prime but no other. We will refer to such a minterm x as an **essential point** of p.

The interesting and computationally attractive point about essential primes is that they can all be determined without having to compute the set of all primes. Two different procedures for determining essential primes have appeared in the literature recently due to Bahnsen, [BAH 81] and Sasao [SAS 83e]. The basic notion used is that an essential prime p must contain a minterm not contained in any other cube $c \subseteq \mathcal{f}$, $c \not\subseteq p$. Bahnsen's method is stated in terms of a single output logic function of Boolean (two-valued) inputs while Sasao deals with the more general case of a multiple-valued input, multiple-output Boolean function. Bahnsen uses the off-set $\mathcal{R}$ and "surrounds" the given prime with cubes $\mathcal{R}'$ in the off-set which are distance 1 from p. He then tests if there exists a minterm in p which is completely "surrounded" by $\mathcal{R}' \cup \{p\}$. In this case p is essential. Sasao surrounds p with distance-1 cubes in $\mathcal{F} \cup \mathcal{D} - \{p\}$ (recall $\mathcal{D}$ is the don't-care set) and tests if the consensus of these with p completely covers p; in which case p is not essential. We use a modification of Sasao's method. Since we have only Boolean inputs, we can use a more efficient data structure than that of Sasao for the general multiple-valued input function, and we can also eliminate some operations required in the general case of multiple-valued inputs. Also, to check if a prime p is covered we use the fast tautology algorithm discussed in Section 4.2.

The problem addressed here is, given a cover

$$\mathcal{F} = \{c^1, \ldots, c^k\}$$

of a multiple output logic function $\mathcal{f}$ where each c^j is a prime cube, determine if a given prime $c^i \in \mathcal{F}$ is an essential prime of $\mathcal{f}$. Since an essential prime must be included in any prime cover, we can determine the set of all essential primes of $\mathcal{f}$ by successively applying the essential prime test to each $c^i \in \mathcal{F}$. If the test for being essential is fast enough, then it is worthwhile to find this set since it will save in computations later on during the main iteration of ESPRESSO-II.

Let $\mathcal{E}$ be the set of all essential primes of $\mathcal{ff}$, and let $e = b(\mathcal{E})$ be the Boolean function represented by $\mathcal{E}$. We use the set $\mathcal{E}$ in the minimization procedure by taking advantage of the following observation:

> If $\mathcal{G}$ is a minimum cover of $(f \cap \bar{e}, d \cup e, r)$, then
> $\mathcal{G} \cup \mathcal{E}$ is a minimum cover of $\mathcal{ff}$.

Thus, knowledge of $\mathcal{E}$ allows us to treat an equivalent problem which has been reduced in size by $|\mathcal{E}|$. More precisely, the number of cubes in the on-set $\mathcal{F}$ is decreased by $|\mathcal{E}|$ and the number of cubes of the don't-care set is increased by the same amount. In our experience, (cf., Chapters 6 and 7) the overall gain in speed is positive and, on the average, enough to justify the overhead in finding $\mathcal{E}$. The set $\mathcal{E}$ is computed by the procedure ESSENTIAL__PRIMES, presented in Figure 4.4.1. The basic ideas of the algorithm implemented in ESSENTIAL__PRIMES are outlined by the following lemmas and theorem.

Lemma 4.4.1: The cube c^i is not essential if and only if for all minterms $x \in c^i$, there exists a minterm $y \notin c^i$, such that $\delta(x,y) = 1$ and $y \in f \cup d$.

Proof: Suppose c^i is not essential; then for any minterm $x \in c^i$ there exists a prime $p \neq c^i$ such that $x \in p$. Consider the cube $p \cap c^i$ and let j be chosen so that $(I(p \cap c^i))_j \neq (I(p))_j$ which exists because $p \neq c^i$ and both are prime. Construct minterm y so that $y_i = x_i$ for all $i \neq j$ and $y_j = 0$ if $x_j = 1$; $y_j = 1$ if $x_j = 0$. Then $\delta(x,y) = 1$, and $y \notin c^i$, $y \in p$; hence $y \in f \cup d$.

Next, suppose for all minterms $x \in c^i$ there exist a minterm $y \notin c^i$, $\delta(x,y) = 1$ and $y \in f \cup d$. Since $x \cup y$ is a cube, there exists some prime $p \supseteq x \cup y$. Since $y \notin c^i$, then $p \neq c^i$; hence for each $x \in c^i$ there is a prime other than c^i containing x, so c^i is not essential. ■

```
Procedure ESSENTIAL__PRIMES(𝓕, 𝓓)

/* Given 𝓕, a prime cover of 𝒻 = (f,d,r)
/* and 𝓓 a cover of d, returns the set of
/* essential primes, 𝓔.
   Begin
   𝓔 ← φ
   for (c^i ∈𝓕)
        Begin
        𝓗̃ ← φ
        for (g ∈ (𝓓∪𝓕)−{c^i})
             Begin
             𝓗̃ ← 𝓗̃ ∪CONSENSUS (g, c^i)
             End
        T←COVERS (𝓗̃,c^i)
        If (T≠1) 𝓔←𝓔∪{c^i}
        End
   Return (𝓔)

   End.
```

Figure 4.4.1

The basic idea of ESSENTIAL__PRIMES is to form the consensus of the cubes in the cover $\mathcal{F}$ with the cube being tested for essentiality, and then to check whether the set of cubes so obtained contains c^i. If it does, then c^i is not essential; otherwise c^i is essential. To justify this procedure, we need the following lemma.

Lemma 4.4.2. Let x,y be minterms such that $\delta(x,y) = 1$. Then, the consensus $p = a \odot b$ of any two cubes $a \supseteq x$ and $b \supseteq y$, with $a \not\supseteq y$, $b \not\supseteq x$, must contain x and y.

Proof: Since the Lemma is symmetric in x and y, we will prove it only for x. Note that $\delta(a,b) \leq 1$, but because $b \not\supseteq x$ and $a \not\supseteq y$, then in fact $\delta(a,b) = 1$. Assume first that $\delta(I(x), I(y)) = 1$ and let k be the conflicting part. Then $x_i = y_i$, $\forall i \neq k$ and the conflicting part of a and b must be k as well, and $a_k = x_k$. Let $p' \subseteq p$ be defined as follows

$$I(p') = \{a_1 \cap b_1, \ldots, a_k, \ldots, a_n \cap b_n\};\ O(p') = O(a) \cap O(b)$$

Since $a \supset x$, $b \supset y$ and $\delta(x,y) = 1$, $a_i \cap b_i \supseteq x_i$ $\forall i, i \neq k$. Hence $x \subseteq p' \subset p$, as claimed.

If $\delta(O(x), O(y)) = 1$, then $\delta(O(a), O(b)) = 1$ since $I(x) = I(y)$ and $a \supset x$, $y \supset b$ implies $\delta(I(a), I(b)) = 0$. Let $p' \subseteq p$ be defined by

$$I(p') = I(a) \cap I(b);\ O(p') = O(a)$$

Then $I(a) \supseteq I(x) = I(y) \subseteq (b)$, so $I(x) \subseteq I(p')$, and of course $O(x) \subseteq O(a)$, so $x \subseteq p' \subseteq p$ as claimed. ■

Theorem 4.4.3 below guarantees that ESSENTIAL_PRIMES is correct.

Theorem 4.4.3. Suppose $\mathscr{F}$ is minimal with respect to single cube containment. Let $\mathscr{H} = (\mathscr{F} \cup \mathscr{D}) - \{c^i\}$, and let $\tilde{\mathscr{H}}$ be the set of cubes obtained as the consensus of $\mathscr{H}$ and c^i:

$$\tilde{\mathscr{H}} = \{h \odot c^i \mid h \in \mathscr{H}\}.$$

Then c^i is covered by $\tilde{\mathscr{H}}$ if and only if c^i is not essential.

Proof. Only if part. Assume $\tilde{\mathscr{H}}$ covers c^i and let x be a minterm in c^i. Then $x \in h \odot c^i$ for some $h \in \mathscr{H}$. If $\delta(c^i, h) = 0$, then $x \in h \cap c^i$ and since $h \not\supseteq c^i$, x is not an essential point of c^i. If $\delta(c^i, h) = 1$, there is a minterm $y \in h$ with $\delta(c^i, y) = 1$, so by Lemma 4.4.2, $y \in h \odot c^i \subseteq \mathscr{F} \cup \mathscr{D}$. Then $x \in h \odot c^i \not\subseteq c^i$ (since $y \notin c^i$) so again x is not an essential point of c^i. But x was an arbitrary minterm of c^i, so c^i is not essential.

If part. Assume c^i is not essential, and x is a minterm of c^i. Then there is a minterm y of $\mathscr{F} \cup \mathscr{D}$ with $y \notin c^i$ and $\delta(x,y) = 1$. Then $y \in h$ for some $h \in \mathscr{H}$, so by Lemma 4.4.2, $x \in h \odot c^i$. Since this is true for any minterm x in c^i, $\tilde{\mathscr{H}}$ covers c^i. ■

To see how ESSENTIAL_PRIMES works, consider the following 3-input, 2-output prime cover

$$\mathcal{F} = \begin{matrix} 2\,0\,0 & 4\,4 \\ 1\,1\,2 & 4\,3 \\ 0\,2\,1 & 4\,3 \\ 0\,1\,2 & 3\,4 \end{matrix},$$

which is illustrated in Figure 4.4.2. Assume there is no don't-care set. Then for $c^1 = 200\ 44$, we find

$$\tilde{\mathcal{H}} = \begin{matrix} 1\,2\,0 & 4\,3 \\ 0\,0\,2 & 4\,3 \\ 0\,2\,0 & 3\,4. \end{matrix}$$

$\tilde{\mathcal{H}}$ does not cover the minterm $100\ 34 \subseteq c^1$, so c^1 is essential. In Figure 4.4.2 the cover $\mathcal{F} = \{c^1,c^2,c^3,c^4\}$ is illustrated, with the essential vertex of c^1 marked with an *. All other primes are not essential. For $c^2 = 112\ 43$, we find

$$\tilde{\mathcal{H}} = \begin{matrix} 1\,2\,0 & 4\,3 \\ 2\,1\,1 & 4\,3 \end{matrix}$$

$(c^2 \odot c^4 = \phi)$ and this clearly covers the two minterms of c^2. The cube c^3 is similar, while for $c^4 = 012\ 34$ we find

$$\tilde{\mathcal{H}} = \begin{matrix} 0\,2\,0 & 3\,4 \\ 0\,1\,1 & 4\,4 \end{matrix}$$

$(c^2 \odot c^4 = \phi)$. Again c^4 is not essential since $\tilde{\mathcal{H}}$ covers it. Note that $c^3 \odot c^4$ covers part of c^4 even though c^3 and c^4 have disjoint output parts.

In ESSENTIAL_PRIMES, the test whether $\tilde{\mathcal{H}}$ covers c^i (procedure COVERS) is performed by using the fast tautology algorithm discussed in Section 4.2 and the result of Proposition 3.1.2 relating tautology and cube covering. Note that the computation time of ESSENTIAL_PRIMES is completely dominated by the tautology test, since the other operations are trivial. The association of essential prime identification with the new fast tautology algorithm presented in Section 4.2 makes the extraction of the essential primes feasible, and the effort expended is generally small compared to the subsequent savings (cf., Chapter 6).

The following result describes another criterion for detecting inessential primes. This result shows that during the EXPAND process, we can sometimes detect such primes with no additional effort.

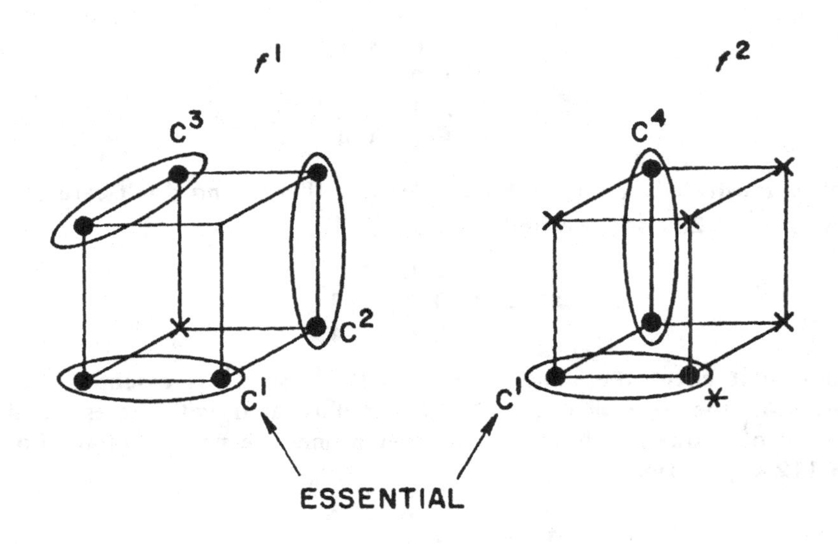

Figure 4.4.2

Recall the notation c^* for the "overexpanded cube" of c, the cube obtained by simultaneously raising all the parts of c which can be individually raised. Equivalently, c^* is the smallest cube containing all the primes which contain c.

Proposition 4.4.4. Suppose $c \in \mathcal{F}$ and p is a prime containing c. If $p \neq c^*$ and p contains no cube of $\mathcal{F}$ other than c, then p is not essential.

Proof: Let x be any minterm of $p \cap \mathcal{F}$; we will show there is a cube $d \subseteq \mathcal{F} \cup \mathcal{D}$ with $x \in d \not\subseteq p$, and hence p is inessential. If $x \notin c$, then $x \in c^i$ for some other cube $c^i \in \mathcal{F}$; but $c^i \not\subseteq p$ so we may take $d = c^i$. But if $x \in c$, just note that $p \neq c^*$ implies there is another prime $q \neq p$ which contains c; so $c \subseteq q \not\subseteq p$ and we may take $d = q$. Therefore p is not essential. ■

Proposition 4.4.4 provides a fast method for detecting some inessential primes during the EXPAND process. Indeed, in the notation of Section 4.3, $c^* = c^+(\mathbb{L}^*, c)$ where $\mathbb{L}^*$ is the set of columns of $\mathbb{B}$

corresponding to row singletons, and $p = c^{+}(\mathbb{L}, c) = c^{*}$ if and only if $\mathbb{L}^{*} = \mathbb{L}$. By the preceding proposition, we can assert that p is inessential if $\mathbb{L} \neq \mathbb{L}^{*}$ and $W = \phi$, where W (computed in EXPAND1) is the set of cubes of $\mathscr{F}$, other than c, covered by p. Such p are flagged during the first pass through EXPAND so we can skip over them during the procedure ESSENTIAL_PRIMES.

Proposition 4.4.4 is based on a related result (Theorem 4) in [HON 74].

Proposition 4.4.5. If $c \in f$ is a minterm and c^{*} is an implicant of $\mathscr{f}$, then c^{*} is an essential prime.

Proof: Since c^{*} contains all the primes containing c, it is itself an implicant of $\mathscr{f}$ only if there is only one prime containing c. This prime is clearly essential and equal to c^{*}. ∎

The converse of this result is also stated in [HON 74]: if c is a minterm in f and c is contained in an essential prime p, then $p = c^{*}$. This is incorrect. It only holds if c is an essential point of p; that is, if c is contained in no prime other than p.

4.5 Irredundant Cover

After the expansion phase of ESPRESSO-II, we have a prime cover $\mathscr{F}$ of $\mathscr{f}$ such that no cube of $\mathscr{F}$ contains any other. However, we have no guarantee that $\mathscr{F}$ is irredundant; it may be that a proper subset of $\mathscr{F}$ is also a cover. Given $\mathscr{F}$ and a cover $\mathscr{D}$ of the don't care set, procedure IRREDUNDANT_COVER produces an irredundant cover $\tilde{\mathscr{F}}$, consisting of a subset of the cubes in $\mathscr{F}$. Consistent with our minimization objective, we try to obtain a subcover $\tilde{\mathscr{F}} \subset \mathscr{F}$ with as few cubes as possible.

Procedure IRREDUNDANT_COVER, shown in Figure 4.5.1, consists of three subprocedures. Procedure REDUNDANT partitions $\mathscr{F}$ into two sets: $\mathbb{E}$ and $\mathbb{R}$. $\mathbb{E}$ is the set of cubes $c \in \mathscr{F}$ such that $\mathscr{F} - \{c\}$ is no longer a cover of $\mathscr{f}$. Clearly such cubes must appear in any subcover $\tilde{\mathscr{F}} \subset \mathscr{F}$; we call them the **relatively essential** cubes of $\mathscr{F}$ (not to be confused with the essential primes discussed in Section 4.4). The remaining cubes $\mathbb{R}$ can each be **individually** removed from $\mathcal{F}$ without destroying the covering property, so we refer to them as the **redundant** cubes of $\mathcal{F}$.

The redundant set $\mathbb{R}$ can be partitioned into the **totally redundant** set $\mathbb{R}_t$ and the **partially redundant** set $\mathbb{R}_p$. The totally redundant set consists of those cubes $c \in \mathbb{R}$ that are covered by $\mathbb{E} \cup \mathcal{D}$. Since $\mathbb{E}$ must be included in any subcover $\tilde{\mathcal{F}} \subset \mathcal{F}$, there is no reason to include these totally redundant cubes, so they are discarded. The remainder of $\mathbb{R}$ forms the partially redundant set $\mathbb{R}_p$. This set is determined by procedure PARTIALLY_REDUNDANT.

Our final (and most complex) task is to extract from $\mathbb{R}_p$ a minimal set $\mathbb{R}_c$ such that $\tilde{\mathcal{F}} = \mathbb{E} \cup \mathbb{R}_c$ is still a cover of ff. Since we choose $\mathbb{R}_c$ to be minimal, $\tilde{\mathcal{F}}$ is an irredundant cover. We attempt to produce an $\mathbb{R}_c$ of minimum cardinality (so that $\tilde{\mathcal{F}}$ is a minimum subcover of $\mathcal{F}$), but we cannot guarantee this. The selection of $\mathbb{R}_c$ is carried out by procedure MINIMAL_IRREDUNDANT.

Example. Let $\mathcal{F} = \{c^1, c^2, c^3, c^4\}$ be a cover for the onset of a single-output function ff shown in Figure 4.5.2. Then $\mathbb{E} = \{c^1, c^4\}$, $\mathbb{R}_t = \phi$ and $\mathbb{R}_p = \{c^2, c^3\}$. If we consider a different function ff' with cover $\mathcal{F}' = \{c^1, c^2, c^3\}$, then $\mathbb{E} = \{c^1, c^3\}$, $\mathbb{R}_t = \{c^2\}$ and $\mathbb{R}_p = \phi$. In this case $\mathbb{E}$ provides the minimum cover for ff'.

Now we describe the details of the three subprocedures. REDUNDANT, illustrated in Figure 4.5.3 and PARTIALLY_ REDUNDANT, illustrated in Figure 4.5.4, have similar structure. Both rely heavily on the tautology algorithm described in Section 4.2.

Procedure REDUNDANT tests each cube $c \in \mathcal{F}$ to see if $(\mathcal{D} \cup \mathcal{F} - \{c\})_c$ is a tautology. By Proposition 3.1.2, this occurs if and only if c is covered by $\mathcal{D}$ and the cubes of $\mathcal{F}$ other than c itself. If c is so covered, it is clearly redundant; otherwise it belongs to the relatively essential set $\mathbb{E}$

Remark: TAUTOLOGY can be used in a "PRESTO-like" logic minimizer. In fact, the strategy of PRESTO [BRO 81] does not involve the construction of a cover $\mathcal{R}$ of the off-set of ff. The expansion operation

Procedure IRREDUNDANT__COVER ($\mathscr{F}$, $\mathscr{D}$)
/* Given $\mathscr{F}$, a cover of $\mathscr{f} = \{f,d,r\}$ and $\mathscr{D}$,
/* a cover of d, return a minimal irredundant cover $\tilde{\mathscr{F}}$.

Begin
$(\mathbb{E}, \mathbb{R}) \leftarrow$ REDUNDANT ($\mathscr{F}$, $\mathscr{D}$)
$\mathbb{R}_p \leftarrow$ PARTIALLY__REDUNDANT ($\mathbb{E}$, $\mathbb{R}$, $\mathscr{F}$)
$\mathbb{R}_c \leftarrow$ MINIMAL__IRREDUNDANT ($\mathbb{R}_p$, $\mathbb{E}$, $\mathscr{D}$)
return ($\tilde{\mathscr{F}} \leftarrow \mathbb{E} \cup \mathbb{R}_c$)
End

Figure 4.5.1

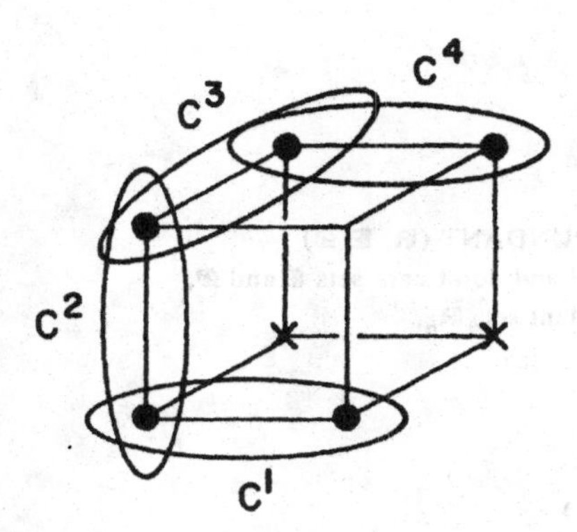

Figure 4.5.2 A Cover with Essential and Redundant Cubes.

```
Procedure REDUNDANT (𝓕, 𝓓)
/* Given 𝓕, a cover of 𝒻 = {f,d,r} and 𝓓, a cover of d,
/* returns 𝔼, the set of relatively essential cubes, and
/* ℝ, the set of redundant cubes of 𝓕.

    Begin
    𝔼 ← φ
    ℝ ← φ
    for (c = 𝓕^1, 𝓕^2,... 𝓕^|𝓕|)
        Begin
        𝓒 ← (𝓕 ∪ 𝓓) − {c}
        if (TAUTOLOGY(𝓒_c)) then ℝ ← ℝ ∪ {c}      /* If cofactor 𝓒_c ≡ 1,
        else 𝔼 ← 𝔼 ∪ {c}                          /* then 𝓒 covers cube c.
        End
    return (ℝ, 𝔼)
    End
```

Figure 4.5.3

```
Procedure PARTIALLY_REDUNDANT (ℝ, 𝔼, 𝓓)
/* Given the relatively essential and don't care sets 𝔼 and 𝓓,
/* returns the partially irredundant set, ℝ_p.

    Begin
    ℝ_p ← φ
    𝓒 ← 𝓓 ∪ 𝔼
    for (r = ℝ^1, ℝ^2,..., ℝ^|ℝ|)
        Begin
        if (1 ≠ TAUTOLOGY (𝓒_r)) ℝ_p ← ℝ_p ∪ {r}    /* if 𝓒_r ≢ 1, r ⊄ 𝓒 so r is partially
                                                   /* redundant (else totally redundant).
        End
    return (ℝ_p)
    End
```

Figure 4.5.4

in PRESTO is performed by asking if a cube $c \in \mathcal{F}$, where $\mathcal{F}$ is the current cover of ff, expanded along a particular coordinate direction, resulting in an expanded cube c^+, is covered by $\mathcal{F} \cup \mathcal{D}$. This is answered by TAUTOLOGY $((\mathcal{F} \cup \mathcal{D})_c)$. ESPRESSO-I implemented the PRESTO strategy by using TAUTOLOGY as described in [BRA 82a].

We now turn to the procedure for determining the partially redundant set $\mathbb{R}_p$ (Figure 4.5.4). A cube c is partially redundant if and only if it is not covered by $\mathbb{E} \cup \mathcal{D}$, and again this can be tested with TAUTOLOGY$((\mathbb{E} \cup \mathcal{D})_c)$.

MINIMAL_IRREDUNDANT is the most complex subprocedure of IRREDUNDANT_COVER. Procedure MINIMAL_IRREDUNDANT returns $\mathbb{R}_c$, a minimal (and hopefully close to minimum) subset of the partially irredundant cubes, such that $\mathbb{E} \cup \mathbb{R}_c$ forms a cover. The minimal irredundant set is computed with the help of an auxiliary single-output logic function $g: B^p \to B$, where $p = |\mathbb{R}_p|$. For $y = [y_1, ..., y_p] \in B^p$, the logic function g is defined by

$$g(y) = 1 \text{ if } \mathcal{F}' = \mathbb{E} \cup \{r^i \in \mathbb{R}_p \mid y_i = 1\} \text{ is a cover of } \mathit{ff}. \quad (4.5.2)$$

In other words, g is 1 only when the nonzero components of y identify a set of cubes in $\mathbb{R}_p$ which, together with $\mathbb{E}$, form a cover. We claim g is a unate function; in fact:

Proposition 4.5.3: The logic function $g: B^p \to B$ defined by (4.5.2) is monotonically increasing in all its input variables.

Proof: Changing an input variable, say y_i, from a 0 to a 1 corresponds to adding the additional cube r^i to the candidate cover $\mathcal{F}'$. In doing so g either stays constant or changes from 0 to 1, since adding yet another cube can never change a cover to a non-cover. Therefore g is monotone increasing in each y_i. ■

The problem of finding the minimum irredundant set is equivalent to that of finding the minimum number of input variables y_i which, when set equal to 1, cause g(y) to be 1. It is not hard to see that this is equivalent to finding the "largest" prime implicant of g.

Now if we could construct a unate cover $\mathscr{G}$ of g, finding the largest prime would be trivial by Proposition 3.3.7: it would simply be the largest cube in the cover. Unfortunately, finding such a cover is rather difficult. It is much easier to obtain a unate cover $\overline{\mathscr{G}}$ of $\bar{g}$, the complement of g, by modifying TAUTOLOGY, as we shall see later. Once such a cover is found, UNATE_COMPLEMENT could be used in principle, to obtain a unate cover $\mathscr{G}$ of g. In some cases, however, the complementation operation is too expensive to be practical, so we use a heuristic algorithm to obtain an approximation to the largest prime in $\mathscr{G}$. This amounts to obtaining an approximate solution to a "simple" column-covering problem.

The following theorem identifies a way of computing a unate cover $\overline{\mathscr{G}}$ of $\bar{g}$, the complement of g.

Theorem 4.5.4. For each $r \in \mathbb{R}_p$, let $\mathscr{P}(r)$ denote the set of minimal sets S such that $(\mathbb{R}_p - S) \cup \mathbb{E} \cup \mathscr{D}$ does not cover r. For each $S \in \mathscr{P}(r)$ define the cube c(S) by

$$c_i(S) = \begin{cases} 0 & r^i \in S \\ 2 & r^i \notin S. \end{cases}$$

Then

$$\overline{\mathscr{G}} = \bigcup_{r \in \mathbb{R}_p} \; \bigcup_{S \in \mathscr{P}(r)} c(S)$$

is a cover of $\bar{g}$.

Proof: Let $y = [y_1, \ldots, y_p]$ be a minterm of $\bar{g}$. Let $S' = \{r^i \in \mathbb{R}_p \mid y_i = 0\}$. Then $(\mathbb{R}_p - S') \cup \mathbb{E}$ is not a cover of $\mathscr{f}$. Equivalently, there is a cube $\bar{r} \in \mathbb{R}_p$ which is not covered by $(\mathbb{R}_p - S') \cup \mathbb{E} \cup \mathscr{D}$. Let $S \subseteq S'$ be a minimal set such that $(\mathbb{R}_p - S) \cup \mathbb{E} \cup \mathscr{D}$ does not cover $\bar{r}$. Then $S \in \mathscr{P}(\bar{r})$ and by definition, y is contained by c(S).

Conversely, let y be a vertex of c(S) for some $S \in \mathscr{P}(r)$, $r \in \mathbb{R}_p$. Then, $S' = \{r^i \in \mathbb{R}_p \mid y_i = 0\} \supseteq S$ and $(\mathbb{R}_p - S') \cup \mathbb{E} \cup \mathscr{D} \subseteq (\mathbb{R}_p - S) \cup \mathbb{E} \cup \mathscr{D}$ does not cover $r \in \mathbb{R}_p$. Hence $\bar{g}(y) = 1$ and the theorem is proved. ∎

Our goal is to find the largest prime in the complement of $\overline{\mathscr{G}}$. By Proposition 4.3.1, this is equivalent to finding the minimum column cover of the matrix $\mathbb{B}(\overline{\mathscr{G}})$ where $\overline{\mathscr{G}} = \{\overline{\mathscr{G}}^i\}$ and

$$\overline{\mathbb{B}}_{ij} = \begin{cases} 1 & \text{if } \overline{\mathscr{G}}^i_j = 0 \\ 0 & \text{if } \overline{\mathscr{G}}^i_j = 2. \end{cases}$$

In fact, a set of cubes $R \subseteq \mathbb{R}_p$, has the property that $R \cup \mathbb{E}$ covers $\mathscr{f}$ if and only if the columns of $\overline{\mathbb{B}}$ corresponding to R cover every row of $\overline{\mathbb{B}}$. Indeed, if $R \cup \mathbb{E}$ is not a cover then $S' = \mathbb{R}_p - R$ contains a minimal set S such that $(\mathbb{R}_p - S) \cup \mathbb{E}$ is not a cover, and since $R \cap S = \phi$, the columns corresponding to R contain no 1 in the row corresponding to c(S). Conversely, if the columns corresponding to R cover $\overline{\mathbb{B}}$, then $R \cap S \neq \phi$ for every such minimal S and hence $R \cup \mathbb{E}$ must be a cover of $\mathscr{f}$.

Therefore the problem of obtaining a minimum cover $\tilde{\mathscr{F}} = R \cup \mathbb{E} \subset \mathscr{F}$ is exactly equivalent to that of finding a minimum column cover of $\overline{\mathbb{B}}$.

To calculate $\overline{\mathbb{B}}$, we need to determine $\mathscr{S}(r)$ for each cube $r \in \mathbb{R}_p$. Now $S \in \mathscr{S}(r)$ if and only if S is a minimal subset of $\mathbb{R}_p$ such that $(\mathbb{R}_p - S) \cup \mathbb{E} \cup \mathscr{D}$ does not cover r. Equivalently, S is a minimal set such that $((\mathbb{R}_p - S) \cup \mathbb{E} \cup \mathscr{D})_r$ is not a tautology. The calculation of such S is carried out by a modification of the TAUTOLOGY procedure. Namely, given covers $\mathscr{A}$ and $\mathscr{B}$ (corresponding to $(\mathbb{R}_p)_r$ and $(\mathbb{E} \cup \mathscr{D})_r$) such that $\mathscr{A} \cup \mathscr{B} \equiv 1$, we determine all minimal subsets $S \subset \mathscr{A}$ such that $(\mathscr{A} - S) \cup \mathscr{B} \not\equiv 1$.

To find such S we apply the unate paradigm. For simplicity suppose $\mathscr{A} \cup \mathscr{B}$ is a single output function. If $\mathscr{A} \cup \mathscr{B}$ is a unate cover, with $\mathscr{A} \cup \mathscr{B} \equiv 1$, then there must be at least one tautologous cube (consisting of all 2's) in the cover $\mathscr{A} \cup \mathscr{B}$. If no such cube occurs in $\mathscr{B}$, take S to be the set of all such cubes; then $S \subset \mathscr{A}$, $(\mathscr{A} - S) \cup \mathscr{B} \not\equiv 1$ and S is the unique minimal subset of $\mathscr{A}$ with this property. On the other hand, if some cube of all 2's appears in $\mathscr{B}$ (so $\mathscr{B}$ itself is a tautology), then we may take $S = \phi$.

To carry out the merge step of the unate paradigm, just note that $(\mathscr{A} - S) \cup \mathscr{B} \not\equiv 1 \Longleftrightarrow (\mathscr{A}_x - S_x) \cup \mathscr{B}_x \not\equiv 1$ or $(\mathscr{A}_{\bar{x}} - S_{\bar{x}}) \cup \mathscr{B}_{\bar{x}} \not\equiv 1$. So to merge the results from each cofactor, we simply take the union of the lists of sets S obtained from each one. Since an S obtained from one cofactor may properly contain an S obtained from the other, some non-minimal S may occur in the new list. But in any case all **minimal** S occur in the list, so the non-minimal ones can be eliminated by inspection later. We can also use the UNATE_REDUCTION algorithm (cf. Theorem 4.2.3) to make the process more efficient.

This algorithm (for the case of a single output function) is presented as procedure LTAUT1 (Figure 4.5.5). LTAUT1 is different from TAUTOLOGY in several respects.

1) It recurs until $\mathscr{B}$ is a tautology or until $\mathscr{A} \cup \mathscr{B}$ is unate. These are the only terminating conditions used.
2) A single-output unate cover is a tautology if and only if it has cubes of all 2's (Proposition 3.3.4).
3) A map $\mathscr{I}$ is maintained which keeps track of the correspondence between the cubes in the recursively cofactored subcover $\mathscr{A}$, and the original cubes in $\mathbb{R}_p$.
4) Undocumented subprocedures SPLIT, and MERGE are called. See comments for the function of these simple routines.

The ideas are illustrated in the following example. Suppose $\mathscr{A}$ is the nine-cube single-output cover given below, and $\mathscr{B} = \phi$. Here $\mathscr{A}$ is a tautology.

$$\begin{array}{c} \\ \\ \\ \\ \mathscr{A} = \\ \\ \\ \\ \\ \end{array}
\begin{array}{cccccc} 1&2&3&4&5&6 \\ \\ 1&0&1&1&0&1 \\ 2&0&1&0&1&0 \\ 1&0&2&1&0&1 \\ 2&2&2&0&2&2 \\ 2&2&2&1&2&2 \\ 2&2&2&2&0&2 \\ 2&2&2&2&1&2 \\ 2&2&2&2&2&0 \\ 2&2&2&2&2&1 \end{array}
\; ; \; \mathscr{I} = \begin{array}{c} (1) \\ (2) \\ (3) \\ (4) \\ (5) \\ (6) \\ (7) \\ (8) \\ (9) \end{array}
\; ; \; \mathscr{A}' = \begin{array}{ccc} 4&5&6 \\ \\ 0&2&2 \\ 1&2&2 \\ 2&0&2 \\ 2&1&2 \\ 2&2&0 \\ 2&2&1 \end{array}
\; ; \; \mathscr{I}' = \begin{array}{c} (4) \\ (5) \\ (6) \\ (7) \\ (8) \\ (9) \end{array}$$

In the unate reduction step (using Theorem 4.2.1), we find $\mathscr{A}$ is a tautology if and only if the lower right-hand block $\mathscr{A}'$ is a tautology. So all minimal sets S will occur among the cubes corresponding to $\mathscr{A}'$.

```
Procedure LTAUT1 (𝒜, ℬ)

/* Given 𝒜, ℬ, covers such that 𝒜∪ℬ ≡ 1,
/* returns B̄, a matrix whose rows mask subsets S ⊆ 𝒜
/* such that (𝒜−S)∪ℬ ≢ 1, and which includes
/* all minimal such S. (Single output version).

    Begin
    if((𝒜∪ℬ) is unate) then
        Begin
        B̄←(row of 0's of length |𝒜|)
        if (ℬ has a row of all 2's) return (B̄)  /* If ℬ ≡ 1, S = φ is
                                                 /* the unique minimal S.
        I←{i: 𝒜^i is a cube of all 2's}          /* Otherwise, S is the
                                                 /* set of cubes in 𝒜 which
                                                 /* are themselves tautologies.
        B̄(I) ← 1
        Return (B̄)
        End
    else
        Begin
        j←BINATE__SELECT(𝒜, ℬ)                  /* Split along x_j
        (𝒮+, 𝒮−)←SPLIT(𝒜, j)                    /* Keep track of cube location
        B̄+ ←LTAUT1(𝒜_xj, ℬ_x̄j)                  /* Work on each cofactor

        B̄− ←LTAUT1(𝒜_xj, ℬ_x̄j)

        B̄←MERGE(𝒮+, 𝒮−, B̄+, B̄−)                 /* Stack the matrices, taking
                                                 /* care to identify columns
                                                 /* to corresponding cubes.
        Return (B̄)
        End
    End
```

Figure 4.5.5

$S^1 = S^{1+} \cup S^{1-}$

$$\mathscr{A}' = \begin{matrix} 0 & 2 & 2 & (4) \\ 1 & 2 & 2 & (5) \\ 2 & 0 & 2 & (6) \\ 2 & 1 & 2 & (7) \\ 2 & 2 & 0 & (8) \\ 2 & 2 & 0 & (9) \end{matrix}$$

$S^{1+} = S^{2+} \cup S^{2-}$

$S^{1-} = \{\{4,6,8\}, \{4,6,9\}, \{4,7,8\}, \{4,7,9\}\}$

x_1

$$\mathscr{A}'_{x_1} = \begin{matrix} 2 & 2 & 2 & (5) \\ 2 & 0 & 2 & (6) \\ 2 & 1 & 2 & (7) \\ 2 & 2 & 0 & (8) \\ 2 & 2 & 1 & (9) \end{matrix}$$

$\bar{x}_1$

$$\begin{matrix} 2 & 2 & 2 & (4) \\ 2 & 0 & 2 & (6) \\ 2 & 1 & 2 & (7) \\ 2 & 2 & 0 & (8) \\ 2 & 2 & 1 & (9) \end{matrix}$$

$S^{2+} = \{\{5,7,8\}, \{5,7,9\}\}$

x_2

$$\begin{matrix} 2 & 2 & 2 & (5) \\ 2 & 2 & 2 & (7) \\ 2 & 2 & 0 & (8) \\ 2 & 2 & 1 & (9) \end{matrix}$$

$S^{2-} = \{\{5,6,8\}, \{5,6,9\}\}$

$\bar{x}_2$

$$\begin{matrix} 2 & 2 & 2 & (5) \\ 2 & 2 & 2 & (6) \\ 2 & 2 & 0 & (8) \\ 2 & 2 & 1 & (9) \end{matrix}$$

x_3

$$\begin{matrix} 2 & 2 & 2 & (5) \\ 2 & 2 & 2 & (7) \\ 2 & 2 & 2 & (9) \end{matrix}$$

$\bar{x}_3$

$$\begin{matrix} 2 & 2 & 2 & (5) \\ 2 & 2 & 2 & (7) \\ 2 & 2 & 2 & (8) \end{matrix}$$

Figure 4.5.6 Example of Generation of all Minimal Non-Covering Sets

We reduce our attention to this new cover; the index sets $\mathscr{I}$ and $\mathscr{I}'$ keep track of the original location of the cubes. Since $\mathscr{A}'$ has no unate columns, we split successively on x_4, x_5 and x_6. The resulting recursion tree is illustrated in Figure 4.5.6. At the top of the figure we see $\mathscr{A}'$ being split into its cofactors with respect to x_1. At the bottom left we see that $\mathscr{A}'_{x_1x_2x_3} \not\equiv 1$ if the cubes $S^{3+} = \{\mathscr{A}^5, \mathscr{A}^7, \mathscr{A}^9\}$ are removed from $\mathscr{A}$. Since $\mathscr{A}'_{x_1x_2x_3} \not\equiv 1$ implies $\mathscr{A}' \not\equiv 1$ and hence $\mathscr{A} \not\equiv 1$, S^{3+} is one of the minimal sets we are seeking. The remaining branches of the computation are only partially shown. In the illustration we have used S^{2+}, S^{2-}, etc. to list sets of cube **indices** instead of the corresponding sets of cubes S. The indices of all minimal sets obtained are accumulated in S^1 = {{5, 7, 8}, {5, 7, 9}, {5, 6, 8}, {5, 6, 9}, {4, 6, 8}, {4, 6, 9}, {4, 7, 8}, {4, 7, 9}}. The Boolean Matrix $\overline{\mathbb{B}}$ associated with this set of "tautology defeating" subsets is thus

$$\overline{\mathbb{B}}(\mathcal{A}) = \begin{array}{ccccccccc} 1 & 2 & 3 & 4 & 5 & 6 & 7 & 8 & 9 \\ \\ 0 & 0 & 0 & 0 & 1 & 0 & 1 & 1 & 0 \\ 0 & 0 & 0 & 0 & 1 & 0 & 1 & 0 & 1 \\ 0 & 0 & 0 & 0 & 1 & 1 & 0 & 1 & 0 \\ 0 & 0 & 0 & 0 & 1 & 1 & 0 & 0 & 1 \\ 0 & 0 & 0 & 1 & 0 & 1 & 0 & 1 & 0 \\ 0 & 0 & 0 & 1 & 0 & 1 & 0 & 0 & 1 \\ 0 & 0 & 0 & 1 & 0 & 0 & 1 & 1 & 0 \\ 0 & 0 & 0 & 1 & 0 & 0 & 1 & 0 & 1 \end{array}$$

Note that $\overline{\mathbb{B}}(\mathcal{A})$ has 1 "Boolean selector" row for each of the 8 "tautology defeating" subsets.

Procedure MINIMAL_IRREDUNDANT is shown in Figure 4.5.7, and calls Procedure NOCOVERMAT which is the subprocedure that builds the matrix $\overline{\mathbb{B}}$, as described in Figure 4.5.8 below. The final step in MINIMAL_IRREDUNDANT is to find the minimum column cover of $\overline{\mathbb{B}}$. Unfortunately, this classical covering problem is NP-complete [GAR 80]. To determine an approximate solution, we implemented the heuristic Procedure MINUCOV illustrated in Figure 4.5.10.

NOCOVERMAT uses LTAUT, the multiple-output version of LTAUT1 (discussed earlier) to build the matrix $\overline{\mathbb{B}}$. For each $r \in \mathbb{R}_p$, LTAUT returns a matrix listing all minimal subsets $S \subset \mathbb{R}_p$ such that $(\mathbb{R}_p - S) \cup \mathbb{E} \cup \mathcal{D}$ does not cover r. NOCOVERMAT simply accumulates the union of these lists by stacking the associated matrices.

To illustrate how NOCOVERMAT works, consider the following covers $\mathcal{F}$ and $\mathcal{D}$ of the onset and don't care set of a multiple output function ff.

```
Procedure MINIMAL__IRREDUNDANT (IR_p, IE, 𝒟)

/* Given IE, the relatively essential set, 𝒟, the don't care set, and
/* IR_p, the partially redundant set, returns IR_c,
/* an approximation to the smallest cardinality subset
/* of IR_p that must be added to IE to form a cover of ℱ.

    Begin
    B̄ ← NOCOVERMAT (IR_p, IE, 𝒟)
    J ← MINUCOV (B̄)
    return IR_c ← {r^j ∈ IR_p | j ∈ J}.
    End
```

Figure 4.5.7

```
Procedure NOCOVERMAT (IE, 𝒟, IR_p)

/* Given IE, the relatively essential set, 𝒟, the don't care set
/* and IR_p, the partially redundant set
/* returns B̄, a Boolean representation of a cover 𝒢̄ of g̃

    Begin
    B̄ ← φ
    for (r = IR_p^1, IR_p^2,... IR_p^|IR_p|)
        Begin
        ℬ ← (IE∪𝒟)_r
        𝒜 ← (IR_p)_r
        𝒮 ← DESCENDANT(𝒜, IR_p)    /* 𝒮 specifies the correspondence between
                                   /* cubes of 𝒜 and cubes of IR_p,
                                   /* that is the i-th cube of 𝒜 has
                                   /* been obtained with the cofactor operation
                                   /* from the 𝒮_i-th cube of IR_p.
        B̄ ← B̄, LTAUT ((𝒜), (ℬ))
        End
    return (B̄)
    End
```

Figure 4.5.8

$$\mathscr{F} = \begin{matrix} 0 & 2 & 1 & 4 & 3 \\ 2 & 1 & 1 & 4 & 4 \\ 1 & 1 & 2 & 4 & 3 \\ 1 & 2 & 0 & 4 & 3 \\ 1 & 2 & 1 & 3 & 4 \\ 1 & 0 & 2 & 3 & 4 \\ 2 & 0 & 0 & 3 & 4 \\ 0 & 1 & 2 & 3 & 4 \end{matrix}, \quad \mathscr{D} = \begin{bmatrix} 1 & 1 & 1 & 4 & 4 \\ 0 & 0 & 0 & 4 & 4 \end{bmatrix},$$

The above covers are illustrated in Figure 4.5.9, which gives the cube index (row index in $\mathscr{F}$ matrix given above) and, in parenthesis, an indication whether the cube is relatively essential ($\mathbb{E}$), partially redundant ($\mathbb{R}_p$), or totally redundant ($\mathbb{R}_t$, cubes 2 and 3).

After PARTIALLY_REDUNDANT complets its computation, we have

$$\mathbb{E} = \begin{bmatrix} 0 & 2 & 1 & 4 & 3 \\ 1 & 2 & 0 & 4 & 3 \\ 0 & 1 & 2 & 3 & 4 \end{bmatrix}; \ \mathbb{R}_p = \begin{bmatrix} 1 & 2 & 1 & 3 & 4 \\ 1 & 0 & 2 & 3 & 4 \\ 2 & 0 & 0 & 3 & 4 \end{bmatrix} \begin{matrix} (1) \\ (2); \\ (3) \end{matrix} \ \mathscr{D} = \begin{bmatrix} 1 & 1 & 1 & 4 & 4 \\ 0 & 0 & 0 & 4 & 4 \end{bmatrix}.$$

When NOCOVERMAT is applied, LTAUT is called on $r^1 = [1\ 2\ 1\ 3\ 4]$ first, and

$$\mathscr{A} = (\mathbb{R}_p)_{r^1} = \begin{bmatrix} 2 & 2 & 2 & 4 & 4 \\ 2 & 0 & 2 & 4 & 4 \end{bmatrix} \begin{matrix} (1) \\ (2) \end{matrix}; \quad \mathscr{B} = (\mathbb{E} \cup \mathscr{D})_{r^1} = [2\ \ 1\ \ 2\ \ 4\ \ 4].$$

In LTAUT, BINATE_SELECT selects the only available variable for splitting, x_2. The cofactor generation gives

$$\mathscr{A}_{x^2} = [2\ \ 2\ \ 2\ \ 4\ \ 4]\ \ (1); \mathscr{B}_{x^2} = [2\ \ 2\ \ 2\ \ 4\ \ 4]$$

$$\mathscr{A}_{\bar{x}^2} = \begin{bmatrix} 2 & 2 & 2 & 4 & 4 \\ 2 & 2 & 2 & 4 & 4 \end{bmatrix} \begin{matrix} (1) \\ (2) \end{matrix}; \ \mathscr{B}_{\bar{x}^2} = \phi.$$

One branch of recursion generates $\bar{\mathbb{B}}^+ = \phi$, since we have a row of all 2's in $\mathscr{B}_{x^2}$, while the other branch returns $\bar{\mathbb{B}}^- = [1\ \ 1\ \ 0]$. Thus we have $\bar{\mathbb{B}} = [1\ \ 1\ \ 0]$.

Next, LTAUT is called first on $r^2 = [1\ \ 0\ \ 2\ \ 3\ \ 4]$. We have

$$\mathscr{A} = (\mathbb{R}_p)_{r^2} = \begin{bmatrix} 2 & 2 & 1 & 4 & 4 \\ 2 & 2 & 2 & 4 & 4 \\ 2 & 2 & 0 & 4 & 4 \end{bmatrix} \begin{matrix} (1) \\ (2) \\ (3) \end{matrix}; \ \mathscr{B} = (\mathbb{E} \cup \mathscr{D})_{r^2} = \phi.$$

LTAUT cannot perform any reduction and selects the only available

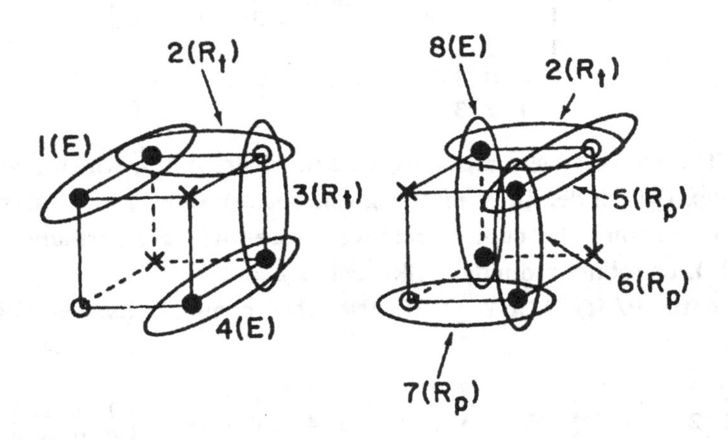

Figure 4.5.9 Cover Example for Procedure NOCOVERMAT

variable for splitting, x_3. The cofactor generation gives

$$\mathcal{A}_{x^3} = \begin{bmatrix} 2 & 2 & 2 & 4 & 4 \\ 2 & 2 & 2 & 4 & 4 \end{bmatrix} \begin{matrix} (1) \\ (2) \end{matrix} ; \mathcal{B}_{x^3} = \phi$$

$$\mathcal{A}_{\bar{x}^3} = \begin{bmatrix} 2 & 2 & 2 & 4 & 4 \\ 2 & 2 & 2 & 4 & 4 \end{bmatrix} \begin{matrix} (2) \\ (3) \end{matrix} ; \mathcal{B}_{\bar{x}^3} = \phi .$$

One branch of the recursion generates $\bar{\mathbb{B}}^+ = [1 \ \ 1 \ \ 0]$ while the second generates $\bar{\mathbb{B}}^- = [0 \ \ 1 \ \ 1]$. We append to $\mathbb{B}$ the matrix

$$\begin{bmatrix} 1 & 1 & 0 \\ 0 & 1 & 1 \end{bmatrix} .$$

Finally, LTAUT is called on $r^3 = [2 \ \ 0 \ \ 0 \ \ 3 \ \ 4]$. We have

$$\mathcal{A} = (\mathbb{R}_p)_{r^3} = \begin{bmatrix} 1 & 2 & 2 & 4 & 4 \\ 2 & 2 & 2 & 4 & 4 \end{bmatrix} \begin{matrix} (2) \\ (3) \end{matrix} ; \mathcal{B} = (\mathbb{E} \cup \mathcal{D})_{r^3} = [0 \ \ 2 \ \ 2 \ \ 4 \ \ 4].$$

Since LTAUT cannot perform any reduction, it selects the only available variable for splitting, x_1. The cofactor generation gives

Procedure MINUCOV ($\bar{\mathbb{B}}$)

```
/* Given B, a Boolean matrix,
/* computes a column cover of B

    Begin
    Ĵ ← φ
    B ← SIMPLIFY (B̄)                    /* Simplify by removing rows that are covered.

    A ← ~B̄ B̄^T                          /* The product of the two matrices is
                                        /* performed by using Boolean operations, and
                                        /* the complement is performed on the
                                        /* result of the product bit by
                                        /* bit on each of the rows.

    Î ← MAXCLIQ(A)                      /* Heuristic procedure to find a large clique,
    for (i ∈ Î)                         /* Î is a large independent set
        Begin
        S ← number of rows of B̄         /* Set of columns in row i
        J ← {j | B̄_ij = 1}
                                        /* Select column covering the most rows
        ĵ ← argmax_{j∈J} ( Σ_{s=1}^{S} B̄_sj )  /* select any of the maximizers
        Ĵ ← Ĵ ∪ {ĵ}                      /* Add this to Ĵ
        B̄ ← REMOVECOV (B, ĵ)            /* Remove column ĵ and
                                        /* all the rows it covers
        End
    if (B̄ ≠ φ) then return (Ĵ ← Ĵ ∪ MINUCOV(B̄))  /* Recur if not done
    else return (Ĵ ← WEED(B, Ĵ)).       /* Reduce to a minimal cover.
```

Figure 4.5.10

$$\mathscr{A}_{x^1} = \begin{bmatrix} 2 & 2 & 2 & 4 & 4 \\ 2 & 2 & 2 & 4 & 4 \end{bmatrix} \begin{matrix} (2) \\ (3) \end{matrix} ; \mathscr{B}_{x^1} = \phi$$

and

$$\mathscr{A}_{\bar{x}^1} = [2\ \ 2\ \ 2\ \ 4\ \ 4]\ (2);\ \mathscr{B}_{\bar{x}^1} = [2\ \ 2\ \ 2\ \ 4\ \ 4].$$

One branch of the recursion generates $\bar{\mathbb{B}}^+ = [0\ \ 1\ \ 1]$, while the second generates $\bar{\mathbb{B}}^- = \phi$, since $\mathscr{B}_{\bar{x}^1}$ has a row of all 2's. The final matrix $\bar{\mathbb{B}}$ constructed by NOCOVERMAT for this example is then

$$\bar{\mathbb{B}} = \begin{matrix} 1 & 1 & 0 \\ 1 & 1 & 0 \\ 0 & 1 & 1 \\ 0 & 1 & 1 \end{matrix}$$

Once $\bar{\mathbb{B}}$ has been obtained, we use the heuristic covering algorithm MINUCOV to find an approximate minimum column cover. MINUCOV is based on an idea of Beister [BEI 80] and a simple greedy strategy.

To find a minimal column cover of $\bar{\mathbb{B}}$, Beister suggests we first find a maximal set I of rows of $\bar{\mathbb{B}}$ such that no two rows in I have a nonzero entry in the same column. Such a set I is called an independent set of rows. Clearly any column cover of $\bar{\mathbb{B}}$ must include at least one column for each row in I, and hence the cardinality of I is a lower bound on the size of a minimum cover. (This fact will be used in Section 6.3 as part of an algorithm to estimate the size of the absolute minimum cover.) Once I has been chosen, we use a greedy strategy to choose a single column covering each row in I. The columns are chosen sequentially. Among our possible choices for a given row in I, we choose the column which covers the most rows in $\bar{\mathbb{B}}$ which have not yet been covered.

If, after choosing a column for each row in I, we still do not have a cover, we delete the columns selected and the rows they cover, and call MINUCOV again. The process continues until a cover is obtained.

Since finding a maximum independent set is again an NP-complete problem, we resort to another heuristic to select a "large" maximal independent set. This is accomplished by building up an auxiliary matrix $\bar{\mathbb{B}}\bar{\mathbb{B}}^T$, where the matrix product is performed with Boolean operators. This square matrix has a 1 in position (i,j) if row i and row j can be covered by a single column. The complement $\sim\bar{\mathbb{B}}\bar{\mathbb{B}}^T$ of this matrix is the adjacency matrix of a graph whose nodes are rows of $\bar{\mathbb{B}}$ and where two nodes are connected by an edge if the corresponding rows are an independent pair. Finding the maximum clique in this graph is equivalent to finding the maximum independent set of rows of $\bar{\mathbb{B}}$ Of course, the maximum clique problem is NP-complete; however, simple greedy heuristics can be used to produce a large clique. This procedure is implemented in MINUCOV as shown in Figure 4.5.10.

Note that we perform a simplification step first. This step identifies any row that "contains" another row in $\bar{\mathbb{B}}$ A row $\bar{b}^i = [\bar{b}^i_1,..., \bar{b}^i_p]$ is contained in $\bar{b}^j = [\bar{b}^j_1,..., \bar{b}^j_p]$ if for all ℓ, $\bar{b}^i_\ell = 1$, implies $\bar{b}^j_\ell = 1$, so any column cover of $\bar{b}^i$ also covers $\bar{b}^j$. Thus $\bar{b}^j$ can be removed.

Note also that it would be useless to search for essential columns of $\bar{\mathbb{B}}$ Such a column would correspond to an element of $\mathbb{R}_p$ that **must** be included in $\mathbb{R}_c$, but since the cubes in $\mathbb{R}_p$ are redundant, any given one can be discarded, so there is no such column.

The procedure we have just described does not necessarily produce a minimal cover. The last step in MINUCOV is the procedure WEED, which reduces the column cover to a minimal set.

Procedure WEED is again a straightforward greedy algorithm. If every column $j \in J$ contains an essential point (a row covered by no other $j' \in J$), then J is minimal. Otherwise, some column has no essential point; among all such columns, we delete from j the one whose removal creates the fewest number of new essential points. The process continues until J is minimal.

Example. Consider the matrix

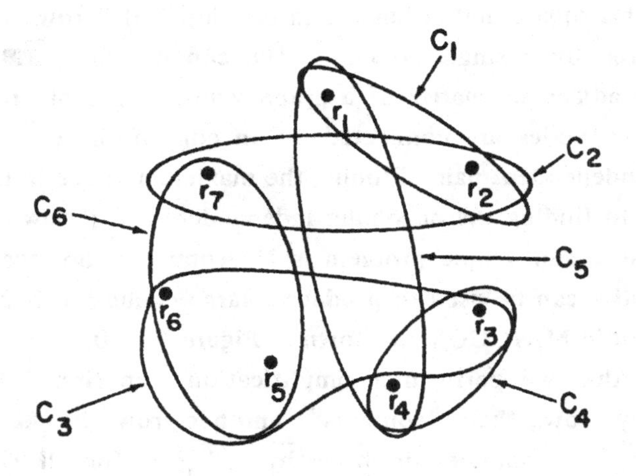

Figure 4.5.11 Covering Problem Example for Procedure WEED.

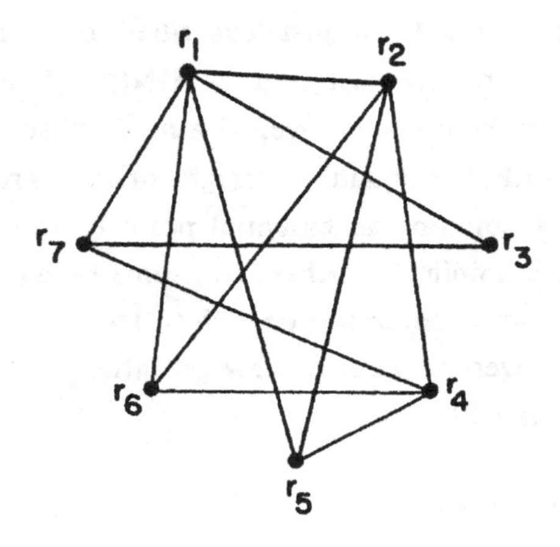

Figure 4.5.12 Independence Graph

$$\overline{\mathbb{B}} = \begin{array}{cccccc|l} c_1 & c_2 & c_3 & c_4 & c_5 & c_6 & \\ 1 & 0 & 0 & 0 & 1 & 0 & r_1 \\ 1 & 1 & 0 & 0 & 0 & 0 & r_2 \\ 0 & 0 & 1 & 1 & 0 & 0 & r_3 \\ 0 & 0 & 0 & 1 & 1 & 0 & r_4 \\ 0 & 0 & 1 & 0 & 0 & 1 & r_5 \\ 0 & 0 & 1 & 0 & 0 & 1 & r_6 \\ 0 & 1 & 0 & 0 & 0 & 1 & r_7 \end{array}$$

representing the 7 point cover problem pictured in Figure 4.5.11. The matrix $\sim\overline{\mathbb{B}}\overline{\mathbb{B}}^T$ represents the graph shown in Figure 4.5.12. From this graph MAXCLIQ selects the independent set $\{r_1,r_3,r_7\}$ which forms a clique. The greedy algorithm then selects columns $\{c_1,c_3,c_6\}$ to cover these rows. The row r_4 is still not covered, so the column c_4 is added to the set. Now the cover is not minimal, the element c_3 is discarded by WEED, and a minimal (in fact minimum) cover $\{c_1,c_4,c_6\}$ is obtained.

We also experimented with another algorithm to find the largest prime of $\mathcal{G}$ and thus the **minimum** irredundant subcover of $\mathcal{F}$. This procedure is a branch and bound modification of UNATE_COMPLEMENT. The procedure is started with a large prime found by the heuristic procedure MINUCOV. The heuristic procedure using MINUCOV is much faster than the branch and bound procedure and we found it usually yields equivalent results. Therefore MINUCOV is the default choice.

Remark: The approach followed in building MINIMAL_IRREDUNDANT has some ideas in common with the one proposed by Rensch [REN 75], Ghazala [GHA 56], and Tison [TIS 67] for generating an irredundant set of prime implicants. Rensch uses an auxiliary logic function as we do and finds a minimal set using a covering algorithm applied to the matrix $\overline{\mathbb{B}}$. However, to the best of our knowledge, both our computational procedure used in building up $\overline{\mathbb{B}}$, which is based on the unate paradigm, and the computation of the minimum irredundant set with a modified version of UNATE_COMPLEMENT are new. The authors mentioned above did not give a systematic algorithm for generating $\overline{\mathbb{B}}$ and did not implement their procedures in a computer program.

4.6 Reduction.

Reduction is an operation that transforms a prime cover $\mathcal{F}$ into a new (in general, not prime) cover $\tilde{\mathcal{F}}$, by replacing each cube by a (usually smaller) cube contained in it.

Reduction allows ESPRESSO-II to move away from a locally optimal solution (an irredundant prime cover is indeed a locally optimal solution) towards a better one. In fact, since some of the cubes of $\tilde{\mathcal{F}}$ are not prime anymore, EXPAND can be applied to $\tilde{\mathcal{F}}$ to yield a different prime cover that may have fewer cubes than $\mathcal{F}$ (never more, since $|\tilde{\mathcal{F}}| \leq |\mathcal{F}|$).

The choice of the cubes to be reduced and the method of reduction have a crucial effect on solution improvement. Our heuristic, similar to the one used in MINI [HON 74], is to process all the cubes of $\mathcal{F}$ sequentially, maximally reducing each one, i.e., making each one as small as possible without destroying coverage.

Remark: If reduction is applied to a cover which is not irredundant, it will generate an irredundant cover, since the redundant cubes will be reduced to empty cubes (which are discarded). MINI does not have an irredundant cover routine and uses reduction to generate irredundant covers. We tried this strategy also, but the use of a specialized irredundant cover routine yields better results according to our experiments.

We define in this section a unate recursive procedure for reduction. We chose this form of REDUCE because the computation for a unate cover can be made by inspection, as will be shown in Section 4.6.3. Our treatment is for the multiple-output case, although the examples treat only the single-output case.

Let $\mathcal{F}$ be a cover of ff, $\mathcal{D}$ a cover of the don't-care set and let c^i be a cube in $\mathcal{F}$. Since we want to maximally reduce c^i while retaining a cover, we would like to find the smallest cube $\underline{c}^i$ containing all the minterms in c^i not covered by $\mathcal{D}$ and $\mathcal{F}-\{c^i\}$. Formally, define $\mathcal{F}(i) \equiv (\mathcal{F}-\{c^i\}) \cup \mathcal{D}$, let $\overline{\mathcal{F}}(i)$ denote the complement of $\mathcal{F}(i)$, and let

$$\underline{c}^i = \underline{\text{S}}\text{mallest } \underline{\text{C}}\text{ube } \underline{\text{C}}\text{ontaining } (c^i \cap \overline{\mathcal{F}}(i)) = \text{SCC}(c^i \cap \overline{\mathcal{F}}(i)).$$

Clearly the smallest cube containing $c^i \cap \overline{\mathcal{F}}(i)$ exists, because it is just the intersection of **all** cubes containing $c^i \cap \overline{\mathcal{F}}(i)$.

The operation performed to obtain $\underline{c}^i$ is called the reduction of c with respect to the cover $\mathcal{F}$. Note that $(\mathcal{F}-\{c^i\}) \cup \{\underline{c}^i\}$ is still a cover of $\mathcal{f}$.

For example, suppose $\mathcal{F} = x_1 + x_2 + x_3$, $c^1 = x_1$. Then $\mathcal{F}(1) = x_2 + x_3$, $\overline{\mathcal{F}}(1) = \bar{x}_2\bar{x}_3$, $c^1 \cap \overline{\mathcal{F}}(1) = x_1\,\bar{x}_2\,\bar{x}_3$, and reduction of $c^1 = x_1$ is $\underline{c}^1 = x_1\bar{x}_2\bar{x}_3$ = the smallest cube containing $x_1\bar{x}_2\bar{x}_3$. Note that $x_1\bar{x}_2\bar{x}_3 + x_2 + x_3 = x_1 + x_2 + x_3$, so the coverage of $\mathcal{F}$ is unchanged by the reduction.

Our procedure for computing $\underline{c}^i$ uses the following identities, which are proved in Lemmas 4.6.1 through 4.6.4 below:

$$\underline{c}^i = c^i \cap (\underline{c}^i)_{c^i}, \tag{4.6.1a}$$

$$(\underline{c}^i)_{c^i} = \mathrm{SCC}(\overline{\mathcal{F}}(i)_{c^i}) = \mathrm{SCCC}(\mathcal{F}(i)_{c^i}), \tag{4.6.1b}$$

where SCCC($\mathcal{G}$)≡<u>S</u>mallest <u>C</u>ube <u>C</u>ontaining <u>C</u>omplement of a given cover $\mathcal{G}$, and

$$\mathrm{SCCC}(\mathcal{G}) = \mathrm{SCC}(x^j\,\mathrm{SCCC}(\mathcal{G}_{x^j}) + \bar{x}^j\,\mathrm{SCCC}(\mathcal{G}_{\bar{x}^j})) \tag{4.6.1c}$$

for any cover $\mathcal{G}$. The procedure REDUCE given in Figure 4.6.1 for reducing all the cubes in a given cover applies the unate recursive paradigm and (4.6.1c) to compute SCCC($\mathcal{G}$).

Reduction as implemented in REDUCE is order dependent. We use a crude *static* reordering strategy. Roughly we choose the largest cube as a seed cube and order the remaining cubes in increasing "pseudo-distance" from this one, where pseudo-distance is measured by the number of mismatches between the two cubes. For example, 012134 and 021144 have a "pseudo-distance" of 3. The heuristic behind this procedure is that the largest cube can be reduced most easily. If we then reduce those cubes which are nearest to it, we increase the probability that later expansion of the large cube will be in the right direction to cover its neighbors. Since the primary goal of our

```
Procedure REDUCE (𝓕, 𝓓)

/* Given 𝓕, a minimal prime cover of
/* 𝒻 = {f,d,r}, and 𝓓, a cover of d,
/* returns the cover 𝓕 where each cube
/* c ∈ 𝓕 is replaced by c̲ = SCCC(𝓗_c).
/* 𝓗 = c∩((𝓕−{c})∪𝓓).

    𝓕←CUBE_ORDER(𝓕)           /* Order cubes to enhance future
                               /* containment by other cubes.
    for (i = 1,..., |𝓕|)
        Begin
        c←𝓕^i
        𝓗←((𝓕−{c})∪𝓓) ∩ c     /* Cover of cubes containing one
                               /* or more minterms of c.
        if (𝓗≠φ)
            Begin
            c̲←c ∩ SCCC(𝓗_c)   /* Reduce c if it intersects cover
            𝓕←(𝓕−{c}) ∪ c̲     /* Update cover.
            End
        End
    return (𝓕)                 /* Return updated reduced cover.
    End
```

Figure 4.6.1

EXPAND algorithm is to cover cubes, it is likely that the reduced seed cube will expand in the right direction. This strategy is embodied in the CUBE_ORDER subprocedure of REDUCE.

Note that in Figure 4.6.1 the body of the **for** loop of REDUCE is skipped if $c \in \mathcal{F}$ intersects no cube in $\mathcal{F} \cup \mathcal{D}$ except itself. If it does intersect, the result (4.6.1b) is applied by calling subprocedure SCCC($\mathcal{H}_c$). In this case c is replaced by $\underline{c}$ in $\mathcal{F}$. REDUCE returns the reduced cover composed of all $\underline{c}^i$.

The remainder of this section is devoted to establishing the unate recursive paradigm for the subprocedure SCCC($\mathcal{H}$).

4.6.1 The Unate Recursive Paradigm for Reduction.

Note that (4.6.1a,b) reduce the REDUCE operation to the problem of evaluating the function SCCC($\mathcal{G}$), where initially $\mathcal{G} \equiv \mathcal{F}(i)_{c^i}$. Equation (4.6.1c) establishes the recursive step. If $\mathcal{G}$ is non-unate, the most binate splitting variable, x_j, is chosen and the problem of evaluating SCCC($\mathcal{G}$) is reduced to evaluating the same function for the smaller covers $\mathcal{G}_{x^j}$ and $\mathcal{G}_{\bar{x}^j}$. If $\mathcal{G}$ is a unate cover, then a special function, SCCCU($\mathcal{G}$), is called which exploits the unateness of $\mathcal{G}$ to find the smallest cube containing its complement. These ideas are embodied in the procedure SCCC shown in Figure 4.6.2. Because of the role of complementation in the functions SCCC and SCCCU, this algorithm closely resembles the unate recursive algorithm for complementation presented in Section 4.1.1.

As an example of the procedure of Figure 4.6.1, suppose $\mathcal{F} \equiv \bar{x}_4 + x_1x_3 + x_1x_2 + \bar{x}_1\bar{x}_2 + \bar{x}_1x_3$, $\mathcal{D} \equiv \phi$ and $c^i = c^1 = \mathcal{F}^1 = \bar{x}_4$. Then $\mathcal{F}(i) \equiv (\mathcal{F} \cup \mathcal{D}) - \{c^i\} = x_1x_3 + x_1x_2 + \bar{x}_1\bar{x}_2 + \bar{x}_1x_3$, and $\mathcal{F}(i)_{c^i} = \mathcal{F}(i)$. Variable x_1 is the most binate so we select $j=1$, and set $\mathcal{G} = \mathcal{F}(i)_{c^i} = \mathcal{F}(i)$. Thus $\mathcal{G}_{x_1} = x_3 + x_2$, $\mathcal{G}_{\bar{x}_1} = \bar{x}_2 + x_3$, which are unate (a fact we ignore for the moment). We note that $\overline{\mathcal{G}}_{x_1} = \bar{x}_2\bar{x}_3$, $\overline{\mathcal{G}}_{\bar{x}_1} = x_2\bar{x}_3$, so SCCC($\mathcal{G}_{x_1}$) = "smallest cube containing $\overline{\mathcal{G}}_{x_1}$" = $\overline{\mathcal{G}}_{x_1} = \bar{x}_2\bar{x}_3$ (because the cover $\overline{\mathcal{G}}_{x_1}$ is a single cube). Similarly, SCCC($\mathcal{G}_{\bar{x}_1}$) = $x_2\bar{x}_3$. Finally, we use (4.6.1c) to obtain $(\underline{c}^i)_{c^i}$ = SCCC($\mathcal{G}$) = SCC($x_1\bar{x}_2\bar{x}_3 + \bar{x}_1x_2\bar{x}_3$) $\equiv \bar{x}_3$. This last identity is obvious geometrically on the Boolean 3-cube but is covered in general by Lemma 4.6.1 below.

Note that procedure SCCC($\mathcal{F}$) is recursive. It calls itself, R_MERGE, BINATE_SELECT and the special function SCCCU for unate $\mathcal{F}$. SCCCU is discussed in detail in Section 4.6.3.

The R_MERGE procedure (which involves merging the two cubes $a = x^j$ SCCC($\mathcal{F}(i)_{x_j}$) and $b = \bar{x}^j$ SCCC($\mathcal{F}(i)_{\bar{x}_j}$) is covered by the following simple Lemma, whose proof is omitted.

Lemma 4.6.1 (R_MERGE): Let a and b be two cubes. Then the smallest cube c containing {a, b} is $c = a \sqcup b$, where $\sqcup$ denotes the coordinatewise union of the entries of the cubes; i.e.

```
Procedure SCCC(ℋ)

/* Given a cover ℋ, returns the smallest
/* cube containing ℋ̄, the complement of ℋ,
/* computed with unate recursive paradigm

    Begin

    If (ℋ unate) then c ← SCCCU(ℋ) else     /* Unate leaf; else:
        Begin
        j ← BINATE_SELECT (ℋ)                /* Select most binate variable.
        c⁺ ← SCCC(ℋ_{x_j})
        c⁻ ← SCCC(ℋ_{x̄_j})
        c ← R_MERGE(x^j c⁺, x̄^j c⁻)          /* Merge according to Lemma (4.6.1)
        End
    return (c)                                /* Smallest cube containing ℋ̄.
    End
```

Figure 4.6.2

	⊔	0	1	2	3	4	φ
		b					
a	0	0	2	2	–	–	0
	1	2	1	2	–	–	1
	2	2	2	2	–	–	2
	3	–	–	–	3	4	3
	4	–	–	–	4	4	4
	φ	0	1	2	3	4	φ

R_MERGE simply applies the table above to the cubes $a = x^j\ SCCC(\mathscr{F}_{x_j})$ and $b = \bar{x}^j\ SCCC(\mathscr{F}_{\bar{x}_j})$.

For example, let $a = 0121343$, $b = 1120434$. Then $c = a \sqcup b = 2122444$. Note that $a \subseteq c$ and $b \subseteq c$, but if any of the raised entries in c were lowered, this would not be true.

4.6.2 Establishing the Recursive Paradigm.

Before describing the special function SCCCU, we prove the basic results (4.6.1) which lead to the unate recursive paradigm for reduction.

Lemma 4.6.2:

$$\underline{c}^i = c^i \cap \mathrm{SCCC}(\mathscr{F}(i)_{c^i}). \tag{4.6.2}$$

Proof: Since $\underline{c}^i \subseteq c^i$, $\underline{c}^i \equiv c^i \cap (\underline{c}^i)_{c^i}$, and by definition

$$(\underline{c}^i)_{c^i} \equiv (\mathrm{SCC}(c^i \cap \overline{\mathscr{F}}(i)))_{c^i}.$$

But since "smallest cube" containing and "cofactor" operations commute, we have

$$(\underline{c}^i)_{c^i} \equiv \mathrm{SCC}((c^i \cap \overline{\mathscr{F}}(i))_{c^i}),$$

and due to the restriction of $(c^i \cap \overline{\mathscr{F}}(i))$ to c^i, the intersection with c^i is irrelevant, so

$$(\underline{c}^i)_{c^i} \equiv \mathrm{SCC}(\overline{\mathscr{F}}(i)_{c^i}).$$

Since by (3.1.4b) $(\overline{\mathscr{G}})_d \equiv \overline{(\mathscr{G}_d)}$ for any cube d and cover $\mathscr{G}$, we have

$$(\underline{c}^i)_{c^i} \equiv \mathrm{SCCC}(\mathscr{F}(i)_{c^i}). \qquad \blacksquare$$

Lemma 4.6.2 enables us to reduce the problem of computing $\underline{c}^i$ to that of computing $\mathrm{SCCC}(\mathscr{G})$ for some given cover $\mathscr{G}$. At the outset, $\mathscr{G} = \mathscr{F}(i)_{c^i}$, but $\mathscr{G}$ changes because we employ, again, the unate recursive paradigm. We establish this recursive paradigm in the present context with Lemma 4.6.3 and Lemma 4.6.4 below.

Lemma 4.6.3: Let $\mathscr{C}$ be a cover of a Boolean function f. Then

$$\mathrm{SCC}(\mathscr{C}) = \mathrm{SCC}(x^j\,\mathrm{SCC}(\mathscr{C}_{x_j}) + \bar{x}^j\,\mathrm{SCC}(\mathscr{C}_{\bar{x}_j})). \tag{4.6.3}$$

Proof: First notice that $SCC(x^j \mathscr{C}_{x_j}) = x^j SCC(\mathscr{C}_{x_j})$ since $\mathscr{C}_{x_j}$ is independent of x^j. Since $\mathscr{C} \supset x^j \mathscr{C}_{x_j}$, we have $SCC(\mathscr{C}) \supset SCC(x^j \mathscr{C}_{x_j}) = x^j SCC(\mathscr{C}_{x_j})$. The same is true for $\bar{x}^j SCC(\mathscr{C}_{\bar{x}_j})$, and therefore

$$SCC(\mathscr{C}) \supset x^j SCC(\mathscr{C}_{x_j}) + \bar{x}^j SCC(\mathscr{C}_{\bar{x}_j}).$$

Applying SCC to both sides and noting $SCC(SCC(\mathscr{C})) = SCC(\mathscr{C})$, we find

$$SCC(\mathscr{C}) \supset SCC(x^j SCC(\mathscr{C}_{x_j}) + \bar{x}^j SCC(\mathscr{C}_{\bar{x}_j})).$$

But $\mathscr{C} = x^j \mathscr{C}_{x_j} + \bar{x}^j \mathscr{C}_{\bar{x}_j} \subset x^j SCC(\mathscr{C}_{x_j}) + \bar{x}^j SCC(\mathscr{C}_{\bar{x}_j})$, so

$$SCC(\mathscr{C}) \subset SCC(x^j SCC(\mathscr{C}_{x_j}) + \bar{x}^j SCC(\mathscr{C}_{\bar{x}_j}))$$

and the proposition follows. ■

Given Lemma 4.6.3 above, it is straightforward to prove Lemma 4.6.4, which directly establishes the basic Shannon recursion (4.6.1c) for the REDUCE operation.

Lemma 4.6.4: Let $\mathscr{G}$ be a cover of cubes. Then

$$SCCC(\mathscr{G}) = SCC(x^j \; SCCC(\mathscr{G}_{x_j}) + \bar{x}^j \; SCCC(\mathscr{G}_{\bar{x}_j})).$$

Proof: Just note that

$$\begin{aligned} SCCC(\mathscr{G}) &= SCC(\overline{\mathscr{G}}) = SCC(x^j SCC(\overline{\mathscr{G}}_{x_j}) + \bar{x}^j SCC(\overline{\mathscr{G}}_{\bar{x}_j})) \\ &= SCC(x^j SCCC(\mathscr{G}_{x_j}) + \bar{x}^j SCCC(\mathscr{G}_{\bar{x}_j})) \end{aligned}$$

where we have used Lemma 4.6.3 and the fact that $(\overline{\mathscr{G}})_{x_j} = \overline{(\mathscr{G}_{x_j})}$. ■

4.6.3 The Unate Case.

In procedure SCCC, if the given cover $\mathscr{F}$ is unate, then instead of "splitting" according to (4.6.1c), we call the special function

SCCCU($\mathcal{F}$). To understand how the unateness property of $\mathcal{F}$ simplifies the computation of SCCC($\mathcal{F}$), consider the following proposition (which is valid in general, not just in the unate case). Recall that u^i is the cube which is a tautology for output i, and ϕ for all other outputs (cf. Section 2.3).

Proposition 4.6.5: The smallest cube c containing the complement of a cover $\mathcal{F}$, $\mathcal{F} \not\equiv 1$, satisfies

$$(I(c))_i = \begin{cases} 0 & \text{if } x^i \subseteq \mathcal{F} \\ 1 & \text{if } \bar{x}^i \subseteq \mathcal{F} \\ 2 & \text{otherwise} \end{cases}$$

$$(O(c))_i = \begin{cases} 3 & \text{if } u^i \subseteq \mathcal{F} \\ 4 & \text{otherwise} \end{cases}$$

Proof: If $x^i \subseteq \mathcal{F}$, then $\overline{\mathcal{F}} \subseteq \bar{x}^i$. Since $\mathcal{F} \not\equiv 1, c_i = 0$. Similarly if $\bar{x}^i \subseteq \mathcal{F}$, then $c_i = 1$. Now suppose that $x^i \not\subseteq \mathcal{F}$ and $\bar{x}^i \not\subseteq \mathcal{F}$. Then there exist minterms z and y such that $z_i = 1$, $z \in \overline{\mathcal{F}}$, $y_i = 0$, $y \in \overline{\mathcal{F}}$; hence $c_i = 2$. If $u^i \subseteq \mathcal{F}$, then $u^i \cap \overline{\mathcal{F}}$ is empty which implies that $(O(c))_i = 3$. Finally, if $u^i \not\subseteq \mathcal{F}$ then there exists a minterm $y \in \overline{\mathcal{F}}$ such that $(O(y))_i = 4$. Thus $(O(c))_i = 4$. ∎

To interpret geometrically the results of Proposition 4.6.5 and their use in computing c = SCCC($\mathcal{F}$), consider the single output case. If $\mathcal{F}$ contains all the vertices of the half space identified by x^i, then the complement contains vertices in the half space identified by $\bar{x}^i$ only. Hence $I(c)_i = 0$. If $\mathcal{F}$ does not contain all the vertices in x^i, then the complement contains vertices in both half spaces and $I(c)_i = 2$.

Proposition 4.6.5 suggests a procedure to construct c. For each input coordinate of c, c_i, $\mathcal{F}$ is tested to see if it contains x^i or $\bar{x}^i$ and the appropriate value of the coordinate of c is selected. For each output coordinate of c, c_j, $\mathcal{F}$ is tested to see if it contains u^j and the appropriate value is selected. This procedure would be very effective if we had a fast way of testing $\mathcal{F}$ for containment of x^i and u^j. Unfortunately, such tests are in general expensive, since we would need to apply procedure COVERS presented in Section 4.3. However, they can be carried out very easily when $\mathcal{F}$ is a unate cover.

Proposition 4.6.6: Let $\mathcal{F}$ be a unate cover. Then $x^i \subseteq \mathcal{F}$ if and only if there exists a subcover $\mathcal{C} = \{c^1, ..., c^m\} \subset \mathcal{F}$ such that $I(x^i) \subseteq I(c^k)$ for all $k = 1,2,..., m$ and $O(x^i) \subseteq O(c^1) \cup ... \cup O(c^m)$.

Proof: Assume $x^i \subseteq \mathcal{F}$. By Proposition 3.3.7, a single-output unate cover contains all the primes of that function, so for each output j there exists a cube $c^j \in \mathcal{F}$ such that $I(x^i) \subseteq I(c^j)$ and $O(u^j) \subseteq O(c^j)$. Taking $\mathcal{C}$ as the set $\{c^1, c^2, ..., c^m\}$, we have $O(x^i) \subseteq O(c^1) \cup ... \cup O(c^m)$ as claimed.

On the other hand, if such a $\mathcal{C}$ exists then obviously $x^i \subseteq \mathcal{F}$. ■

A similar criterion tests if $\bar{x}^i \subseteq \mathcal{F}$. The test for $u^i \subseteq \mathcal{F}$ is even simpler: $u^i \subseteq \mathcal{F}$ if and only if there exists a cube $c \in \mathcal{F}$ such that $I(c) = I(u^i) = 222...2$ and $O(c) \supset O(u^i)$. This follows simply from the fact that a single-output unate tautology must include a cube of all 2's.

These tests are applied directly in Procedure SCCCU shown in Figure 4.6.3.

The following example shows how SCCCU operates. Let

$$M(\mathcal{F}) = \begin{matrix} 2\ 1\ 2 & 4\ 4 \\ 2\ 2\ 1 & 3\ 4 \end{matrix}$$

be the matrix representation of a unate 2-output function. This cover is represented by the cubes c^1 and c^2 shown in Figure 4.6.4. Black dots indicate vertices of the onset, crosses those of the offset. The sets N^1 and $\bar{N}^1$ referred to in the procedure of Figure 4.6.3 are both empty so we set $y_1 = 2$. The set $N^2 = \{4,5\}$, hence $y_2 = 0$. The set $N^3 = \{5\}$ and $\bar{N}^3 = \phi$, hence $y_3 = 2$. Since there are no rows of all twos, $y_4 = y_5 = 4$ hence

$$y = 2\ 0\ 2\ \ 4\ 4$$

is the smallest cube containing the complement of $\mathcal{F}$. This cube is labelled D^1 in Figure 4.6.4.

```
Procedure  SCCCU(𝓕)

/* Given a unate cover 𝓕, returns the smallest
/* cube y containing the complement of 𝓕.
/* The cubes of 𝓕 have n inputs and m outputs.

    Begin
    for (i = 1,..., n)
        Begin
        N^i ← {j | I(x^k) ≤ I(𝓕^k) AND 𝓕_j^k = 4 for some k}   /* Finds the cubes with inputs equal to
                                                                 /* I(x^i) and tags the output of these cubes.
        N̄^i ← {j | I(x̄^i) ≤ I(𝓕^k) AND 𝓕_j^k = 4 for some k}  /* Same as above for x̄^i.

        if ( | N^i | = m) then y_i = 0                          /* 𝓕 contains x^i
        else
            Begin
            if ( | N̄^i | = m) then y_i = 1                      /* 𝓕 contains x̄^i
            else y_i = 2                                        /* 𝓕 does not contain x̄^i or x^i
            End
        End
    for (i = n + 1,..., n + m)
        Begin
        if (I(𝓕^k) = I(u^i) AND 𝓕_i^k = 4) for some k           /* 𝓕 contains u^i
        then y_i = 3 else y_i = 4                               /* 𝓕 does not contain u^i
    return (y)
    End
```

Figure 4.6.3

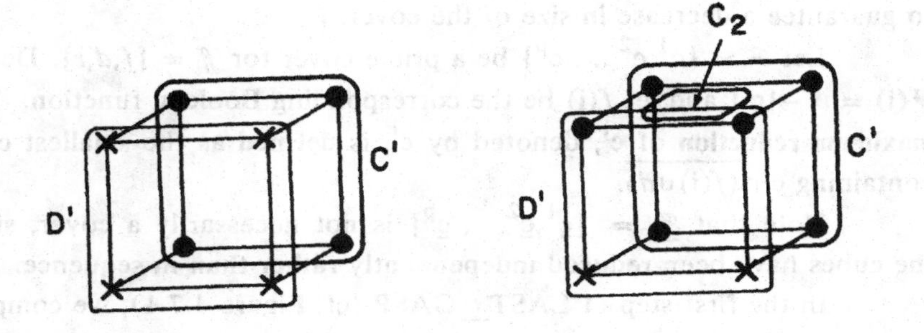

Figure 4.6.4 Cover Example for Illustration of Procedure SCCU

4.7 Lastgasp.

This procedure is used as a final attempt to extract a few more cubes from the cover. The procedure LAST_GASP shown in Figure 4.7.1 is a modified REDUCE followed by a modified EXPAND, and is based on the following ideas. We have seen that REDUCE is an effective strategy for decreasing the size of a cover. In fact, by reducing a cube c, we have two possibilities for decreasing the size of the cover:

i) The reduced cube $\tilde{c}$ can be covered by a neighboring cube after the expansion;

ii) The reduced cube $\tilde{c}$ can expand in different directions to cover some neighboring cube.

In both cases, the best results are obtained when the reduced cube is the smallest. We noted in Section 4.6 that REDUCE is dependent on the order in which the cubes are processed. Cubes occurring late in the ordering have less chance of being reduced, because the cubes preceding them have already been made smaller, and the covering has therefore become "tighter". The order in which the cubes are reduced is determined with a crude heuristic and the entire reduction process does not give any guarantee that the cover size will be decreased by EXPAND. In contrast, LAST_GASP reduces each of the cubes in the cover "maximally", then applies an expansion step on the reduced cubes and keeps track of the coverage of the re-expanded cubes in an attempt to guarantee a decrease in size of the cover.

Let $\mathscr{F} = \{c^1, c^2, ..., c^p\}$ be a prime cover for $\mathscr{f} = \{f, d, r\}$. Define $\mathscr{F}(i) = \mathscr{F} - \{c^i\}$ and let $f(i)$ be the corresponding Boolean function. The **maximum reduction of** c^i, denoted by $\underline{c}^i$, is defined as the smallest cube containing $c^i \cap (f(i) \cup d)$.

Note that $\underline{\mathscr{F}} = \{\underline{c}^1, \underline{c}^2, ..., \underline{c}^p\}$ is not necessarily a cover, since the cubes have been reduced independently rather than in sequence.

In the first step of LAST_GASP (cf. Figure 4.7.1), we compute the set $\underline{\mathscr{F}}$ of maximally reduced cubes described above. Next the set $\underline{\mathscr{F}}$ is restricted to those cubes which succeeded in being reduced. These are expanded as a set; that is, the part of the expand operation (cf.

```
Procedure LAST__GASP(𝓕, 𝓓, 𝓡)
 /* Given a prime cover 𝓕 of ff = {f,d,r}, a cover 𝓓 of d and 𝓡 of r,
/* returns a prime cover 𝓕, of possibly decreased cardinality.

  Begin
     𝓕̲ ← MAXIMUM__REDUCTION(𝓕, 𝓓)
     𝓖 ← EXPAND (𝓕̲ − {c̲_i | c̲_i = c_i}, 𝓡)                /* Expand only those cubes
                                                          /* which have been reduced.
     𝓗 ← {g^i | g^i ∈ 𝓖, g^i ⊃ c̲^k ∈ 𝓕̲ for some k≠i}.        /* Use only additional cubes
                                                          /* which succeeded in covering
                                                          /* a maximally reduced cube.
     if (𝓗 ≠ φ), then 𝓕 ← IRREDUNDANT__COVER(𝓕∪𝓗, 𝓓)
     𝓕 ← IRREDUNDANT__COVER(𝓕∪𝓗, 𝓓)
     return (𝓕)
  End
```

Figure 4.7.1

Section 4.3.2) that guides the expansion towards covering another cube, only looks at cubes of this set. Note that cubes which were not reduced are still prime so they cannot be covered.

The set $\mathcal{H}$, constructed in LAST__GASP, consists of only those expanded cubes g^i which succeeded in covering one of the reduced cubes $\{\underline{c}^j\}$. besides the original cube $\underline{c}^i$ (which is always contained in g^i.) These are exactly the reduced and re-expanded cubes which may reduce the size of the cover. If the reduced and re-expanded cubes are the same as the original cubes in $\mathcal{F}$, we have performed useless operations. However, the following theorem ensures that the cubes in $\mathcal{H}$ are indeed new primes.

Theorem 4.7: Let $\mathcal{F}$ be an irredundant prime cover of ff. Let $\mathcal{H}$ be obtained by LAST__GASP. If $g^i \in \mathcal{H}$, then g^i is prime and $g^i \notin \mathcal{F}$.

Proof: The cube g^i is prime because it is produced by the EXPAND operation (cf. Theorem 4.3.4). Since $g^i \in \mathcal{H}$, it covers at least two reduced cubes, say $\underline{c}^i$ and $\underline{c}^k$, and hence it also covers the essential parts of c^i and c^k. Now if g^i were an element of $\mathcal{F}$, then c^i or c^k would be redundant, contrary to assumption. Therefore $g^i \notin \mathcal{F}$. ∎

If $\mathcal{H}$ is not empty, LAST_GASP adds the new primes in $\mathcal{H}$ to the cover $\mathcal{F}$. There is then a good chance that IRREDUNDANT_COVER can reduce the cover $\mathcal{F} \cup \mathcal{H}$ to an irredundant cover of cardinality strictly less than that of $\mathcal{F}$. In fact, we might be tempted to claim that g^i can always replace at least the two cubes c^i and c^k, whose maximal reductions $\underline{c}^i$ and $\underline{c}^k$ are covered by g^i. Although this has always been the case in our experience, the following counterexample exists.

Let

$$\mathcal{F} = \begin{bmatrix} 1 & 2 & 2 \\ 2 & 2 & 1 \end{bmatrix} \quad \mathcal{D} = \begin{bmatrix} 1 & 0 & 0 \\ 0 & 1 & 0 \\ 0 & 0 & 1 \end{bmatrix}.$$

Then

$$\underline{c}^1 = [1 \ 1 \ 0]$$
$$\underline{c}^2 = [0 \ 1 \ 1]$$

$$\mathcal{H} = \{g^1\} = [2 \ 1 \ 2]$$

$$\mathcal{F} \cup \mathcal{H} = \begin{bmatrix} 1 & 2 & 2 \\ 2 & 2 & 1 \\ 2 & 1 & 2 \end{bmatrix}$$

but no single cube of $\mathcal{F} \cup \mathcal{H}$ is a cover for ff, and hence the size of the cover cannot be reduced to less than the original size of $\mathcal{F}$. Although g^1 covers $\underline{c}^1 \cup \underline{c}^2$, it does not cover the minterm [101] in the on-set. See Figure 4.7.2 below; black dots, white dots and X's indicate the on, off and don't-care vertices respectively. The cube g^1 covers $\underline{c}^1$ and $\underline{c}^2$ but does not cover the entire on-set.

Nevertheless, we reemphasize that we have never experienced a practical case in which $\mathcal{H} \neq \phi$ did not cause the cover to decrease in size.

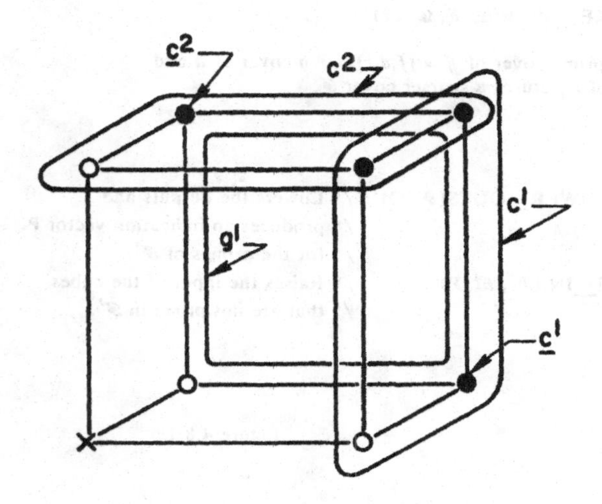

Figure 4.7.2 Maximally Reduced Cubes for LAST__GASP Example.

4.8 Makesparse.

As the name suggests, the procedure MAKE__SPARSE attempts to make the PLA matrix as sparse as possible. This enhances its ability to be folded and improves some of its electrical properties. MAKE__SPARSE, shown in Figure 4.8.1, is a simple two step procedure which operates on an irredundant prime cover $\mathcal{F}$.

The first step (Procedure LOWER__OUTS) lowers as many 4's to 3's as possible in an attempt to make the output plane of the PLA sparse. The second step (Procedure RAISE__IN) makes the input part as sparse as possible by raising 1's and 0's to 2's. Note that if the sequence in which the two procedures are applied is reversed, RAISE__IN would not raise any entry, since the cover $\mathcal{F}$ is already a prime cover. However, if we first lower the outputs, then the cubes of $\mathcal{F}$ that have been changed by LOWER__OUTS are no longer prime and may possibly be expanded in the input part. To save computer time, LOWER__OUTS produces, as in REDUCE, a vector P indicating which

```
Procedure MAKE__SPARSE(𝓕, 𝓓, 𝓡)

/* Given 𝓕 a prime cover of ff = {f,d,r}, 𝓓 a cover of d and
/* 𝓡 a cover of r, returns a sparser cover 𝓕.

    Begin

    (𝓕', P) ← LOWER__OUTS(𝓕, 𝓓)    /* Lowers the outputs and
                                    /* produces an indication vector P
                                    /* for the primes of 𝓕'.
    𝓕 ← RAISE__IN (𝓕', 𝓡, P)        /* Raises the input of the cubes
                                    /* that are not prime in 𝓕'.

    return (𝓕)
    End
```

Figure 4.8.1

```
Procedure LOWER__OUTS(𝓕, 𝓓)

/* Given a prime cover 𝓕 of ff = {f, d, r}, 𝓓 a cover of d,
/* returns a new cover 𝓕, with lowered output and an indicator P to
/* mark the cubes that were not changed; m is the number of outputs of 𝓕

    Begin
    for (i = 1,..., |𝓕|)
       P_i ← 1
    for (k = 1,..., m)
         Begin
         ~𝓕 ← {u^k c^i | c^i ∈ 𝓕}          /* Restrict 𝓕 to output k.
         ~𝓓 ← {u^k d^i | d^i ∈ 𝓓}          /* Restrict 𝓓 to output k.
         O(𝓕)_k ← IRR(~𝓕, ~𝓓)              /* The output column k is replaced
                                           /* by a vector indicating an irredundant
                                           /* cover for output k.

         End
    for (i = 1,..., |𝓕|)
         if (𝓕^i was changed) then P_i ← 0
    return (𝓕, P)
    End
```

Figure 4.8.2

of the $c^i \in \mathscr{F}$ remain prime, so that RAISE_IN will not attempt to raise their inputs.

LOWER_OUTS could be implemented as a modified version of REDUCE, in which only the output parts are reduced, or of IRREDUNDANT_COVER, in which the outputs of ff are considered one at a time. We experimented with both and concluded that the implementation using IRREDUNDANT_COVER was computationally more efficient. This version of LOWER_OUTS is shown in Figure 4.8.2.

We note that at the end of MAKE_SPARSE the new cover $\mathscr{F}$ is not necessarily a prime cover but it is sparse. If we were to iterate the procedure, we could guarantee that the new cover $\mathscr{F}$ would be maximally sparse in the sense that no nontrivial entry in $\mathscr{F}$ (i.e. 0, 1, or 4) could be eliminated (i.e. changed to 2 or 3) without $\mathscr{F}$ ceasing to be a cover. However, most of the time we have run MAKE_SPARSE the cover was maximally sparse after one iteration. Thus we chose not to iterate for the sake of efficiency. It is important that the essential primes $\mathscr{E}$ be added back to the cover before MAKE_SPARSE is executed (cf. Figure 4.0.1), since even essential primes can be made sparse.

In LOWER_OUTS a vector P is constructed indicating if c^i remains prime. $\tilde{\mathscr{F}}$ and $\tilde{\mathscr{D}}$ are the restrictions of $\mathscr{F}$ and $\mathscr{D}$ to the single output k. The procedure IRR uses IRREDUNDANT_COVER and returns an indicator vector giving a minimal subset of $\tilde{\mathscr{F}}$ which is a cover of output k. IRR returns a vector whose i-th entry is 4 if c^i is in the set and is 3 otherwise. This replaces the kth output column of $\mathscr{F}$; output k of $\mathscr{F}$ has then been "lowered". The procedure IRR differs from IRREDUNDANT_COVER in that only one output occurs in the output part.

The procedure RAISE_IN is shown in Figure 4.8.3. It is a modification of EXPAND, restricted to raising only the input part of each cube of $\mathscr{F}$. It uses the prime indicator, P, constructed in LOWER_OUTS to skip over cubes which are already prime. For each cube $c^i \in \mathscr{F}$ that is not prime, the set of cubes $\tilde{\mathscr{R}}$ of $\mathscr{R}$ having at least one output in common with c^i is formed, and the blocking matrix $\mathbb{B}$ described in Section 4.3.1 is constructed using only the input part of c^i

and the input part of $\tilde{\mathscr{R}}$. We then perform EXPAND, and replace the input part of c^i with the prime which results. In this way, the input part of the cube c^i is raised to become as sparse as possible.

```
Procedure RAISE__IN(𝓕, 𝓡, P)

/* Given a cover 𝓕 of 𝒻 = {f,d,r} and a cover 𝓡 of r
/* and a prime indication vector P, returns a cover 𝓕, where
/* each input part of 𝓕 is raised to be a maximum prime.

  Begin
  for (i = 1,..., |𝓕|, and P_i ≠ 1)                 /* Do for each input cube not prime.
        Begin
        𝓡̃ ← {r^j ∈ 𝓡 | O(𝓕^i)_k = O(r^j)_k = 4  for some k}   /* Gather up those cubes of the offset
                                                     /* which effect the outputs of 𝓕^i.
        𝔹 ← BLOCK(I(𝓕^i), I(𝓡̃))                     /* Form blocking matrix.
        I(𝓕^i) ← c⁺(MINLOW(𝔹), 𝓕^i)                  /* Replace input part of 𝓕^i
                                                     /* by a minimal one.
        End
  return (𝓕)
  End
```

Figure 4.8.3

4.9 Output Splitting

As pointed out by Sasao et. al [SAS 84], there exist perfectly reasonable functions whose complements are perfectly unreasonable. A good example is the "Achilles' heel function":

$$f = x_1x_2x_3 + x_4x_5x_6 + \dots + x_{3n-2}x_{3n-1}x_{3n}.$$

This function is unate, has 3n variables, and its minimum cover has n terms. However, $\bar{f}$ which is also unate has a minimum cover of 3^n terms.

Such examples require that logic minimizers, such as ESPRESSO-II and MINI, which form a representation of the complement (note that PRESTO does not have this problem) must be able to detect this situation and develop an alternate strategy. We have implemented in ESPRESSO-II a technique called "output splitting" which

1) estimates the complement size for each single-output function,

2) divides each function with a large complement into smaller functions with reasonable estimated complement size,
3) minimizes the resulting multiple-output function, and
4) merges the components of the split outputs back into a single-output.

Roughly speaking, we partition an unmanageable problem into manageable pieces. Note that during the minimization process, the several outputs, which result from splitting a single output whose complement is too large, are treated as unrelated. Therefore, this technique may give up some optimality in order to handle such large-complement functions.

Note, in addition, that this technique is heuristic, i.e., it will not always reduce the size of the cover of the complement computed by COMPLEMENT. However, in our experience, when this strategy is applied in those situations where the complement has a very large estimated size, it results in sufficient reduction of the complement size so that computation can proceed without undue expenditure of resources.

The two major parts of this strategy are the procedure for estimating the complement size and the method for dividing a cover $\mathcal{F}$ into a union of smaller covers $\mathcal{F} = \mathcal{F}_1 \cup \mathcal{F}_2 \cup \mathcal{F}_3 \cup \ldots \cup \mathcal{F}_n$ such that each $\mathcal{F}_i$ has a reasonable complement size.

We first describe ESTCOMP, the complement size estimator, illustrated in Figure 4.9.1, which is designed for speed. Our strategy for estimating the size of the cover produced by COMPLEMENT, is to find an upper bound for the cardinality of a **minimum** cover of the complement of $\mathcal{F}$, called the **complement size,** $cs(\mathcal{F}) = \min \{ |\mathcal{R}| : \mathcal{R}$ is a cover of the complement of $\mathcal{F}\}$. The estimator is based on the idea of looking for disconnected components in the personality matrix of the cover. The following propositions establish the rules used for estimating $cs(\mathcal{F})$.

Proposition 4.9.1. If $\mathcal{F} = c \cap \mathcal{F}_c$ where c is a cube and $\mathcal{F}_c$ is not empty, then $cs(\mathcal{F}) = cs(c) + cs(\mathcal{F}_c)$.

```
Procedure ESTCOMP(F~)

/* Given F~ = F ∪ D, where F is a cover of ff = {f,d,r}
/* and D is a cover of d, estimates the size of the complement
/* of F and returns N, an upper bound on the minimum
/* complement size.

    Begin
    F~ ← SIMPLIFY_COMPLEMENT(F~)                     /* Remove single variable
                                                     /* cubes and make remaining
                                                     /* cover independent of
                                                     /* these variables.
                                                     /* (Proposition 4.9.4)

    if (|F| = 1) return (VAR_COUNT(F~))              /* Return with the number
                                                     /* of variables in the cube.
                                                     /* (DeMorgans Law)
    (F~_1,...,F~_n) ← COMPONENTS(F~)                 /* Find components of F~.
    if (n≠1)
        Begin
        N ← 1                                        /* (Proposition 4.9.3)
        for (j = 1,..., n), N ← N × ESTCOMP(F~_j)
        return (N)
        End
    c ← CUBE_FACTOR(F~)                              /* Extract any cube
                                                     /* multiplier of F~.
    if (c≠φ)
        return (N ← ESTCOMP(c) + ESTCOMP(F~_c))      /* (Proposition 4.9.1)
    j ← LEAST_TWOS(F~)                               /* Choose the column
                                                     /* with the least number of 2's.
              |F|
    if (0 =   Σ   (F~^i_j = 2)) then
             i=1
        return (N ← ESTCOMP(F~_{x_j}) + ESTCOMP(F~_{x̄_j}))   /* (Proposition 4.9.2)
    else
        Begin
        F~_1 ← {F~^i | F~^i_j ≠ 2}
        F~_2 ← {F~^i | F~^i_j = 2}
        return (N ← ESTCOMP(F~_1) × ESTCOMP(F~_2))   /* (Proposition 4.9.3)
        End
    End
```

Figure 4.9.1

Proof. By De Morgans' law, for any covers $\bar{c}$ of the complement of c and $\overline{\mathscr{F}}_c$ of the complement of $\mathscr{F}_c$, $\mathscr{R}' = \bar{c} \cup \overline{\mathscr{F}}_c$, is a cover of the complement of $\mathscr{F}$. Hence,

$$cs(\mathscr{F}) \leq cs(c) + cs(\mathscr{F}_c).$$

To obtain the opposite inequality, assume for simplicity that $c = x^i$; the general result will follow by induction. Then the complement of $\mathscr{F}$ is $\bar{x}^i + \overline{\mathscr{F}}_c$. Since $\mathscr{F}_c$ is not empty, $\overline{\mathscr{F}}_c$ is not a tautology and hence $\bar{x}^i$ is an essential prime of $\overline{\mathscr{F}}_c$, and hence, $cs(\overline{\mathscr{F}}) \geq 1 + cs(\overline{\mathscr{F}}_c) = cs(c) + cs(\overline{\mathscr{F}}_c)$. ■

Proposition 4.9.2. $cs(\mathscr{F}) \leq cs(\mathscr{F}_{x_j}) + cs(\mathscr{F}_{\bar{x}_j})$.

Proof. By (3.1.8),

$$x^j \overline{\mathscr{F}}_{x_j} + \bar{x}^j \overline{\mathscr{F}}_{\bar{x}_j}$$

is a cover of the complement of $\mathscr{F}$. ■

Remark. Note that $cs(\mathscr{F})$ is not necessarily equal to $cs(\mathscr{F}_{x_j}) + cs(\mathscr{F}_{\bar{x}_j})$, since some cubes in $x^j \hat{\overline{\mathscr{F}}}_{x_j}$, where $\hat{\overline{\mathscr{F}}}_{x_j}$ is a minimum cover of the complement of $\mathscr{F}_{x_j}$, can possibly be combined with cubes in $\bar{x}^j \hat{\overline{\mathscr{F}}}_{\bar{x}_j}$ to form a smaller cover of the complement of $\mathscr{F}$.

Proposition 4.9.3. If $\mathscr{F} = \{\mathscr{F}_1\} \cup \{\mathscr{F}_2\}$ then

$$cs(\mathscr{F}) \leq cs(\mathscr{F}_1) \times cs(\mathscr{F}_2). \tag{4.9.1}$$

Proof. Let $\overline{\mathscr{F}}_1$ and $\overline{\mathscr{F}}_2$ be two minimum covers of the complement of $\mathscr{F}_1$ and $\mathscr{F}_2$. By De Morgan's law

$$\overline{\mathscr{F}} = \{\overline{\mathscr{F}}_1\} \cap \{\overline{\mathscr{F}}_2\}$$

is a cover of the complement of $\mathscr{F}$. The intersection between the two sets of cubes has cardinality less than or equal to the product of the cardinality of the two sets, so the proposition follows. ■

One might suspect that equality holds in (4.9.1) if $\mathcal{F}_1$ and $\mathcal{F}_2$ are independent in some sense. Let us say $\mathcal{F}_1$ and $\mathcal{F}_2$ have **disjoint variable sets** if they depend on distinct sets of variables. More precisely, if some cube in $\mathcal{F}_1$ has a 0 or 1 in input variable position i, then all cubes in $\mathcal{F}_2$ have a 2 in that position, and vice-versa.

Proposition 4.9.4 If $\mathcal{F}_1$ and $\mathcal{F}_2$ have disjoint variable sets, and $\mathcal{F}_1$ consists of a *single cube,* then $cs(\mathcal{F}_1 \cup \mathcal{F}_2) = cs(\mathcal{F}_1) \times cs(\mathcal{F}_2)$.

Proof. Suppose $\mathcal{F}_1 = \{x_1 x_2 ... x_n\}$, and $\overline{\mathcal{F}}$ is a minimum prime cover for the complement of $\mathcal{F}_1 \cup \mathcal{F}_2$. Since $\mathcal{F}_1$ and $\mathcal{F}_2$ are on disjoint variable sets, each cube of $\overline{\mathcal{F}}$ can be factored as $c = \bar{x}_i\, d$ for a **unique** $\bar{x}_i$, $1 \le i \le n$. (There must be one such i since $x_1 ... x_n$ does not meet $\overline{\mathcal{F}}$, and uniqueness is a consequence of the primality of c.) Then for each i, $\overline{\mathcal{F}}_{\bar{x}_i} = \overline{\mathcal{F}}_{x_1 x_2 ... \bar{x}_i ... x_n} = \overline{\mathcal{F}}_2$; since the cubes of $\overline{\mathcal{F}}$ participating in $\overline{\mathcal{F}}_{\bar{x}_i}$ are different for each i, $|\overline{\mathcal{F}}| = cs(\mathcal{F}_1 \cup \mathcal{F}_2) = n \times cs(\mathcal{F}_2)$ $= cs(\mathcal{F}_1) \times cs(\mathcal{F}_2)$. ■

It is a remarkable fact that this proposition **fails** if we do not require that $\mathcal{F}_1$ consists of a single cube. Equivalently, when f and g are two Boolean functions on disjoint variable sets, it is **not** necessarily true that a minimum cover for fg can be obtained as the product of a minimum cover for f and a minimum cover for g. There exist examples where min (fg) < min (f) • min (g), where min(f) is the minimum number of cubes in a cover of f, and similarly for min(g) and min(fg).

To understand why such examples exist, consider a simpler covering problem. Let $A = \{1,2,3\}$, $B = \{4,5,6\}$, and let $\mathcal{A}$ and $\mathcal{B}$ consist of all two-element subsets of A and B. Then $\min(A,\mathcal{A}) = \min(B,\mathcal{B}) = 2$; that is, a minimum covering of A by sets in $\mathcal{A}$ consists of 2 elements, and similarly for $(B,\mathcal{B})$. Now let $\mathcal{A} \times \mathcal{B} = \{a \times b \subseteq A \times B \mid a \in \mathcal{A}, b \in \mathcal{B}\}$. Then $\min(A \times B, \mathcal{A} \times \mathcal{B}) = 3 < 4 = \min(A,\mathcal{A}) \cdot \min(B, \mathcal{B})$. Indeed, a three element cover for $A \times B$ is provided by $\{1,2\} \times \{4,5\}$, $\{2,3\} \times \{5,6\}$ and $\{1,3\} \times \{4,6\}$.

We now describe a disguised version of this example in terms of

Boolean functions. Let

$$f = (A+B+C)D + ABC + A\overline{B}\overline{C} + \overline{A}B\overline{C} + \overline{A}\overline{B}C$$
$$= f_1 + f_2$$

$$g = (E+F+G)H + EFG + E\overline{F}\overline{G} + \overline{E}F\overline{G} + \overline{E}\overline{F}G$$
$$= g_1 + g_2$$

Here f is represented as the sum of all of its prime implicants; one prime is redundant, which we may take to be either AD, BD or CD; and a similar statement holds for g. Thus min (f) min (g) = 6•6 = 36. But

$$fg = f_2g_2 + f_2((E + F)H) + g_2((A + B)D) + DH(AE + BF + CG)$$

exhibits fg as a union of 4•4 + 4•2 + 4•2 + 3 = 35 cubes. Thus min(fg) < min(f) min(g). If one notes that min (f_1) = min $(g_1) = 2$, while DH(AE + BF + CG) covers the "essential part" of f_1g_1 with 3 cubes, it is not too difficult to establish a correspondence between this example and the simpler one preceding it.

We now turn to the application of Propositions 4.9.1 through 4.9.4 to the estimation of $cs(\mathscr{F})$, the size of a minimum cover for the complement of $\mathscr{F}$. We focus on the following situations:

1) cs(c) = number of literals in a cube c. (DeMorgan's Law)
2) $cs(\{x^i\} \cup \mathscr{F}) = cs(\mathscr{F})$ (Proposition 4.9.4 when $\mathscr{F}$ does not depend upon the ith input.)
3) $cs(\mathscr{F}_1 \cup \mathscr{F}_2) \approx cs(\mathscr{F}_1) \times cs(\mathscr{F}_2)$ (Proposition 4.9.3 when $\mathscr{F}_1$ and $\mathscr{F}_2$ are disjoint.)
4) $cs(c \cap \mathscr{F}_c) = cs(c) + cs(\mathscr{F}_c)$ (Proposition 4.9.1.)
5) $cs(\mathscr{F}) \leq cs(\mathscr{F}_{x_j}) + cs(\mathscr{F}_{\overline{x}_j})$ (Proposition 4.9.2.)
6) $cs(\mathscr{F}_1 \cup \mathscr{F}_2) \leq cs(\mathscr{F}_1) \times cs(\mathscr{F}_2)$ (Proposition 4.9.3.)

The procedure ESTCOMP of Figure 4.9.1 is recursive. If any of the special cases represented by relations 1-4 above is found, $\mathscr{F}$ is reduced and ESTCOMP is applied to this. In these cases nearly exact estimates are obtained. (Although 3 in the disjoint variable case is not exact, it is almost always so.) The last two cases will only give upper bounds and are applied when none of the earlier special cases occur. The heuristic for deciding between case 5 or 6, is to find the column in $\mathcal{F}$ with the least number of 2's. If this column has no 2's, then case 5 is used, otherwise case 6 is used. In applying 6, $\mathcal{F}_1$ and $\mathcal{F}_2$ are chosen according to

$$\mathcal{F}_1 = \{\mathcal{F}^i \mid \mathcal{F}^i_j \neq 2\}$$

$$\mathcal{F}_2 = \{\mathcal{F}^i \mid \mathcal{F}^i_j = 2\}.$$

where j is the column with the least 2's. The heuristic is motivated primarily by those cases which have disconnected components or nearly disconnected components. By removing a column with many non 2's, a nearly disconnected cover may become disconnected allowing us to apply Proposition 4.9.3 with near certainty that the estimate will be exact. Because cofactoring is done only on a column with no 2's, the procedure has the property that the total number of cubes examined never exceeds $|\mathcal{F}|$. Therefore there is no storage expansion during recursion. The procedure is also fast because at each leaf, only a number N is estimated and returned, and during the merging on the upward movement in the recursion tree, this is only multiplied or added to the other results.

ESTCOMP merely gives an *upper bound* on the complement size, and in a few cases this bound may be much higher than the true value. However, the purpose of ESTCOMP is not to produce an exact estimate but to detect a potential complement explosion, and for this purpose it functions well.

Next we describe output splitting which is similar to ESTCOMP in many ways since output splitting is applied in those cases where the estimated complement size was large. The splitting is done in a way which eliminates the reasons why the estimate was large. Output splitting is done on any single-output function whose estimated complement size exceeds a threshold value MAXCOMP (default = 1000). The result of the procedure OUTSPLIT is a multiple-output function such that the union of all of its single-output functions is the initial function. In addition, the total number of terms is the same. Thus if $\mathcal{F}$ is a cover of the single output function f, and $\mathcal{G}$ is the result of the splitting operation, then $\mathcal{F} = \mathcal{G}_1 \cup \ldots \mathcal{G}_i \cup \ldots \cup \mathcal{G}_k$ where $\mathcal{G}_i$ is the cover of the ith output of $\mathscr{G}$. Further

$$|\mathscr{F}| = \sum_{i=1}^{k} |\mathscr{G}_i|,$$

so in fact the $\mathscr{G}_i$ form a partition of the cubes of $\mathscr{F}$.

Procedure OUTSPLIT is shown in Figure 4.9.2. Basically, we look for disconnected components of the cubes of a single-output cover $\mathscr{F}$. If only one component is found, we choose a variable with the least number of twos and put its index in a set J. Then the procedure COMPONENTS is recalled, but the variables in the set J are to be ignored. This step is iterated until we have suppressed enough variables to cause more than one component to be formed. When there are several components, OUTSPLIT is called recursively for any component whose estimated complement size exceeds MAXCOMP. The successive calls to OUTSPLIT will then cause a further refinement in the partition. Finally GATHER is called to group together blocks of the partition whose union has an estimated complement size not exceeding MAXCOMP.

Finally, observe that we cannot claim that OUTSPLIT will always succeed in reducing the size of the complement. We can only claim that it will reduce the estimated size of the complement. Further, assuming that the complement algorithm always produces a cover size that is better than or equal to the estimate, we can guarantee that enough space has been allocated before complementation is performed. This procedure produces a reasonably safe (but not guaranteed) approach for the practical examples where complementation can cause trouble. For example, when OUTSPLIT is applied to the Achilles' heel function, the desired effect is produced. For example, each of the n cubes of the Achilles' heel function are identified as a separate disconnected component. Then GATHER forms groups of size k where $3^k \leq$MAXCOMP. Each of these groups then forms a new output satisfying the bounds on its complement size.

```
Procedure OUTSPLIT(F~)

/* Given F~ = F∪D, where F is a cover of ff~= {f,d,r} and
/* D is a cover of d, partitions the cubes of F~ into sets
/* {F~_i} such that ESTCOMP(F~_i) ≤ MAXCOMP.
/* The cubes of F~ which belong to F are tagged to distinguish them
/* from the cubes of D.

    Begin
    J ← φ
    k ← 1
    while (k = 1)
        Begin
        (k,F~_1,F~_2,..., F~_k) ← COMPONENTS(F~,J)  /* Form the disjoint components
                                                   /* of F~ ignoring the variables
                                                   /* in the set J.
        J ← J∪LEAST__TWOS(F~,J)                    /* Select the column not in
                                                   /* J which has the least
                                                   /* number of 2's.
        if (J = {1,..., n})                        /* If we fail to produce
          Begin                                    /* more than one component,
          k ← |F~|                                 /* partition F~ into its
          (F~_1,..., F~_k) ← (F~^1,..., F~^k)      /* individual cubes.
          End
        End
    for (j = 1,..., k)
        Begin
        if (MAXCOMP<ESTCOMP(F~_j)                  /* Keep splitting components
          (F~_{j_1},..., F~_{j_s}) ← OUTSPLIT(F~_j)  /* until each estimate is
                                                   /* small enough.
        End

    return (GATHER({F~_{i_n}})                     /* Form groups of {F~_{i_n}}
                                                   /* whose complement sizes
                                                   /* do not exceed MAXCOMP
    End                                            /* The covers returned by
                                                   /* GATHER are assigned to
                                                   /* F and D according to
                                                   /* the tags assigned to the
                                                   /* cubes of F~ before
                                                   /* entering OUTSPLIT.
```

Figure 4.9.2

Chapter 5

MULTIPLE-VALUED LOGIC MINIMIZATION

5. Multiple-Valued Logic Minimization

As discussed in Chapter 1, many multiple-valued input logic functions can be implemented very efficiently in PLA's with two-bit input decoders. Hence multiple-valued logic minimization is of great practical importance [FLE 75]. A two-bit decoder pairs two Boolean variables, say x_1 and x_2, and generates four decodes

$$x_1 x_2,$$
$$\bar{x}_1 x_2,$$
$$x_1 \bar{x}_2,$$
$$\bar{x}_1 \bar{x}_2.$$

The inverse of these are fed into the input plane of a PLA. These four replace x_1, $\bar{x}_1$, x_2, $\bar{x}_2$ as inputs to the PLA, so the number of input lines remains the same. The input plane of the PLA performs AND operations on its inputs, so with two-bit decoding we can form all fourteen distinct nontrivial functions of x_1, x_2 by AND-ing together the four decodes in various ways, including each of the eight functions that can be generated by AND-ing together the two original Boolean inputs and their complements. In terms of logic minimization and the area of the PLA, the use of two-bit decoders can only result in an improvement; but a penalty must be paid for the additional area and delay introduced by the decoders themselves.

It is obviously important to take advantage of the many values that an "input" can take when allowing bit pairing. In fact, the output of the two-bit decoder can be seen as a single four-valued input variable to the PLA. Multiple-valued input logic minimization attempts to minimize the number of product terms needed to implement the given logic function by exploiting the flexibility offered by the values that the inputs can take. Note also that the results obtained by multiple-valued logic minimization of decoded inputs obviously depend on which inputs are paired together. Several algorithms [SAS 83b] have been proposed for assigning the pairing. We will restrict our attention to that portion of the logic minimization problem for which the multiple-valued inputs are given.

Multiple-valued logic minimization has recently received increased interest due to a connection that has been made between it and the state assignment problem [DEM 83f]. Briefly stated, the idea of this connection is that the states of a finite state machine can be represented as the set of possible values for a *single* multiple-valued variable. The PLA is optimized in this form. This process combines certain subsets of the states in rows of the resulting PLA, and these sets are then used as the input to an imbedding procedure which attempts to assign each of these sets to subcubes of a Boolean k-cube while minimizing k. This procedure can be used also to assign codes to the inputs and outputs of a finite state machine, as well as to solve other encoding problems.

Several techniques can be used to minimize multiple-valued logic functions (e.g. [SU 72]). Most of them are extensions of two-valued logic minimization. In this section, we show how to effectively use ESPRESSO-II (or any PLA minimizer constructed *only* for two-valued logic and which allows a don't-care set), to minimize logic functions whose *inputs* are multiple-valued.

For convenience, we will restrict our attention to single-output functions. A multiple-valued input, single-output logic function $\mathscr{f}$ is a map

$$\mathscr{f}: P \rightarrow \{0,1,2\}$$

where

$$P = \mathop{\mathrm{X}}_{i=1}^{n} P_i$$

and each P_i is a set of integers $\{1,2,\ldots, p_i\}$. An element $v = (v_1,\ldots, v_n)$ of the input space P is just a sequence of integers such that $1 \le v_i \le p_i$; it differs from the usual Boolean input in that each v_i may assume p_i values instead of just two. The numbers $\{1,2,\ldots, p_i\}$ are just labels for the values of the incoming variable; in the case of a two-bit decoder, $\{1,2,3,4\}$ would correspond to the conditions x_1x_2, $x_1\bar{x}_2$, $\bar{x}_1x_2$ and $\bar{x}_1\bar{x}_2$ (only one of which can occur for any fixed value of x_1 and x_2).

For purposes of minimization, we will represent each v_i as a p_i-tuple of numbers such that the v_i-th is 1 and the rest are zero. For example, if $n = 3$, and $(p_1,p_2,p_3) = (3,5,2)$, then $v = (2,4,1)$ will be represented by (0 1 0) (0 0 0 1 0) (1 0). This should be thought of as a "minterm". A general product term might look like:

$$c = (1\ 1\ 0)\ (0\ 1\ 1\ 0\ 1)\ (1\ 0).$$

This cube c represents the condition (v_1 = 1 or 2) **and** (v_2 = 2, 3 or 5) **and** (v_3 = 1). The problem of multi-valued logic minimization is to find an optimal expression for $\mathscr{f}$ as a sum of cubes of the above form. Note that in general a **single** cube can represent a complicated conjunctive form, involving both **ands** and **ors** (product-of-sums).

This representation, introduced in [SU 72], is the one used by MINI. An innovation used in MINI was to note that the output part of a logic function can be treated in the same way as a single multiple-valued input, thus reducing the complexity of the algorithms. Indeed, the output part signifies that the cube appears in each of the outputs whose bits are raised, just as the input part represents the "**or**" of the corresponding input conditions.

In representing multiple-valued variables using two-valued variables, we construct a so-called "one-hot" code for each variable having more than two values. We illustrate this notion for a variable v which can take on three values. To this variable we associate three Boolean variables, x_1, x_2, x_3, such that x_i is 1 exactly when $v = i$. Exactly one of the x_i is 1 for any given value of v, so the possible values of x_1, x_2, x_3 are covered by

$$\begin{array}{ccc} x_1 & x_2 & x_3 \\ \\ 1 & 0 & 0 \\ 0 & 1 & 0 \\ 0 & 0 & 1 \end{array} \quad . \tag{5.1}$$

These are the care vertices on the Boolean 3-cube associated with (x_1, x_2, x_3). The complement of the set above is the don't-care set associated with these variables. In this case the don't-care set is

$$\begin{array}{ccc} 1 & 1 & 2 \\ 1 & 2 & 1 \\ 2 & 1 & 1 \\ 0 & 0 & 0 \end{array} \quad . \tag{5.2}$$

This set simply records the fact that no pair of the x_i are ever **both** on, and the x_i are never **all** off. Since these conditions never occur, they are don't-care conditions.

We have seen previously that a multiple-valued input logic function can represent in a single cube a logical expression involving both **ands** and **ors**. In contrast, a regular logic function with only Boolean (two-valued) inputs can only express a series of **ands** in a single cube. It is appropriate then to ask how we can expect a single cube of Boolean variables to express a product-of-sums. The key is the don't-care set. Consider a single part representing a four-valued variable v and the condition

(1 0 1 0)

i.e., the condition v=1 or v=3. This is represented using Boolean variables x_1, x_2, x_3, x_4 as

2 0 2 0

i.e., **not** x_2 **and not** x_4. Because the don't-care set is specified so that one and only one of the x_i can be on at one time, this is equivalent to specifying (x_1 or x_3), which says exactly (v=1 or v=3). Another way to see this is that there are only two care vertices contained in the cube 2 0 2 0, namely 1 0 0 0 and 0 0 1 0, which label the conditions v=1 and v=3.

This discussion is summarized in the following result.

Proposition 5.1: Let $\varepsilon = (\varepsilon_1, \ldots, \varepsilon_n)$ be a sequence of 0's and 1's. Then the cover $\mathcal{F} = \{x^i \mid \varepsilon_i = 1\}$ and the cube c defined by

$$c_i = \begin{cases} 0 & \text{if } \varepsilon_i = 0 \\ 2 & \text{if } \varepsilon_i = 1 \end{cases}$$

represent the same Boolean function relative to the don't-care set

$$\mathcal{D} = \{x_i x_j \mid i \neq j\} \cup \{\bar{x}_1 \bar{x}_2 \ldots \bar{x}_n\}. \tag{5.3}$$

Proof: We must show that a minterm $x \in B^n - \mathcal{D}$ is in $\mathcal{F}$ if and only if it is in c; i.e., $\mathcal{F} - \mathcal{D} = c - \mathcal{D}$. Such a minterm x must have the form

$$x_i = \begin{cases} 1 & i = j \\ 0 & i \neq j \end{cases}$$

for some j, i.e., $x \in x^j$. Then $x \in \mathcal{F} \Longleftrightarrow \varepsilon_j = 1$, while $x \in c \Longleftrightarrow c_j = 2 \Longleftrightarrow \varepsilon_j = 1$, so $\mathcal{F} - \mathcal{D} = c - \mathcal{D}$ and $\mathcal{F}$ and c represent the same logic function relative to $\mathcal{D}$. ■

It is not hard to see how this result can be used to carry out multiple-valued logic minimization with a Boolean minimizer. First, we create $\Sigma\, p_i$ Boolean variables by associating to each multi-valued input v_i a variable representing the condition $v_i = j$, $j = 1, \ldots, p_i$. The group of Boolean variables associated to a given v_i is called an input **part.** To each part we associate a don't-care set of the form given in (5.3), specifying that exactly one variable in that part is "on" at any given time. The total don't-care set is the union of those for each part. By the above theorem, we can now represent any multi-valued cube, for example

$$(1\ 1\ 0)\ (0\ 1\ 1\ 0\ 1)\ (1\ 0),$$

by an equivalent Boolean cube; in this case,

$$c = 2\,2\,0\quad 0\,2\,2\,0\,2\quad 2\,0.$$

The set of all such cubes is minimized using the don't care set. The result may then be converted back into the multi-valued form.

Even a single-output Boolean minimizer can be applied to a multiple-output, multi-valued input logic function, because the outputs can themselves be treated as a single part. However, each part contributes $O(p_i^2)$ terms to the don't-care set, so, if there are many outputs, it is preferable to use a multiple-output Boolean minimizer. Indeed, if even an input part is very large, the above method will be less efficient than true multi-valued minimization. However, for many purposes - including bit-pairing - the overhead is negligible.

Finally, we describe two examples in which the above procedure was used by ESPRESSO-II for multi-valued minimization. The first example is provided by the PLA shown in Figure 5.1. The don't-care set (specified by (5.3)) is not shown. The first eight variables represent a single 8-valued variable and the remaining input variables represent a single 4-valued variable; there are only two (multi-valued) input variables, and nine outputs. The appropriate don't-care set was provided to ESPRESSO-II and the result of the minimization is shown in Figure 5.2. Using the MINI representation of multiple-valued logic functions the same function is shown in Figure 5.3. This example illustrates how, in the final representation, a single cube can represent a product-of-sums. For example, row 1 of Figure 5.3 is

01001101 0010 | − − 1 − − − − − − 1−

representing for outputs 3 and 10 the condition

$$(v_2^1 + v_5^1 + v_6^1 + v_8^1) \wedge v_3^2$$

where v_2^1 means the second value of the first 8-valued variable. This same example was run using APL MINI in true multi-valued mode and

```
100000001000|10000000---    100000000010|01000000---
001000001000|10000000---    001000000010|01000000---
010000001000|00100000---    010000000010|00100000---
000010001000|00100000---    000010000010|00100000---
000100001000|00010000---    000001000010|00100000---
000000101000|00010000---    000000010010|00100000---
000000011000|00010000---    000100000010|00010000---
000001001000|00000010---    000000100010|00001000---
010000001000|--------000    100000000010|--------001
000010001000|--------000    001000000010|--------001
000001001000|--------000    010000000010|--------010
100000001000|--------001    000100000010|--------010
001000001000|--------001    000010000010|--------010
000000101000|--------010    000001000010|--------010
000100001000|--------100    000000100010|--------010
000000011000|--------100    000000010010|--------010
001000000100|10000000---    001000000001|01000000---
000000100100|10000000---    000000100001|01000000---
100000000100|00010000---    000010000001|00100000---
010000000100|00010000---    000001000001|00100000---
000010000100|00010000---    000000010001|00100000---
000100000100|00001000---    100000000001|00001000---
000000010100|00001000---    000100000001|00001000---
000001000100|00000001---    010000000001|00000100---
010000000100|--------000    010000000001|--------000
000010000100|--------000    100000000001|--------010
000001000100|--------000    000100000001|--------010
100000000100|--------010    000010000001|--------100
000100000100|--------100    000001000001|--------100
000000010100|--------100    000000010001|--------100
001000000100|--------101    001000000001|--------101
000000100100|--------101    000000100001|--------101
```

Figure 5.1 Multi-valued Input Version of PLA DK17 (Don't-cares not shown)

the same result was obtained except for an additional 1 in row 13 column 12 of Figure 5.3.

There is a possible advantage in using ESPRESSO-II to minimize a multi-valued function with the approach outlined in this chapter versus the approach followed by MINI. In MINI, the expansion process is performed by selecting a **part** and by expanding it maximally before choosing another part.

With the ESPRESSO-II approach, each **component** of each part can be expanded independently, so that the resulting prime may be obtained by only partially expanding each part before a component of

```
0.00..0...1.|--1------1
0000..0....1|--1-----1-
.0.00000..1.|-1--------
..00.000.1..|---1------
.0.00000.000|1---------
0.00.000.0.0|--1-------
00.000.0...1|-1------1-
000.000..1..|----1---1-
000.000..000|---1----1-
.00.0000...1|----1----1
00.000.0.1..|1-------1-
.....1...1..|-------1--
.0000000.1..|---------1
......1...1.|----1----1
000.0000..1.|---1-----1
......1..000|---1-----1
.....1...000|------1---
.1.........1|-----1----
```

Figure 5.2 Multi-valued Input Minimization of DK17

```
010011010010|--1------1-
000011010001|--1-----1--
101000000010|-1--------1
110010000100|---1-------
101000001000|1---------1
010010001010|--1--------
001000100001|-1------1-1
000100010100|----1---1--
000100011000|---1----1--
100100000001|----1----1-
001000100100|1-------1-1
000001000100|-------1---
100000000100|---------1-
000000100010|----1----1-
000100000010|---1-----1-
000000101000|---1-----1-
000001001000|------1----
010000000001|-----1-----
```

Figure 5.3 MINI Representation of Minimized DK17

another part is selected for expansion. In other words, the MINI expansion process is limited to only a subset of the set of all primes that cover the given cube. Thus the ESPRESSO-II expansion process has more degrees of freedom for possibly obtaining better results. A positive effect has been observed in practice on a number of problems.

A second example is provided by the 4-bit adder function. We know, [SAS 83b], that by pairing variables (1,5), (2,6), (3,7), (4,8), and converting each pair into a single 4-valued variable, it is possible to obtain a PLA with only 17 terms, which is the provable minimum. When ESPRESSO-II was applied to this problem and provided with the appropriate don't-care set, the minimum cover was indeed achieved.

Chapter 6

EXPERIMENTAL RESULTS

6. Experimental Results.

In this chapter we discuss our computational experience with an APL implementation of ESPRESSO-II. We report data on 56 PLA's given to us by various sources. Some have already been minimized by various PLA minimization programs while others consist entirely of minterms. The 56 PLA's are a mixture of control and data flow logic. Some have don't-care sets and there is a range in size from 4 inputs to 94 inputs, from 1 output to 109 outputs, and from an initial cover size of 10 terms to 654 terms. The computing time required by ESPRESSO-IIAPL to minimize these PLA's ranged from less than a second to 721 seconds on an IBM/3081K machine. Most of the PLA's represent real applications although a few are arithmetic functions which we devised for their interesting structures, or because they were known to be difficult problems.

One goal in this chapter is to give the reader some empirical evidence for degree of optimization that can be achieved by

Name	In	Out	Type	DC	Literals Init	Literals Fin	Terms Init	Terms Fin	Time Sec
ADD6	12	7	D	0	9840	2551	1092	355	118.2
ADR4	8	5	D	0	2672	415	255	75	9.3
ALU1	12	8	D	0	60	60	19	19	.6
ALU2	10	8	D	1	593	347	87	68	12.0
ALU3	10	8	D	1	352	351	68	65	13.4
APLA	10	12	C	1	1277	222	112	25	7.7
BCA	26	46	C	1	5905	3263	301	180	230.9
BCB	26	39	C	1	5684	2765	299	155	125.2
BCC	26	45	C	1	4903	2542	245	138	121.2
BCD	26	38	C	1	4483	2024	243	117	77.8
BCO	26	11	C	0	6673	2084	419	179	101.8
CHKN	29	7	C	0	1865	1739	153	140	50.9
CO14	14	1	C	1	210	210	14	14	.4
CPS	24	109	C	0	7810	2893	654	160	723.9
DC1	4	7	C	0	69	54	15	9	.9
DC2	8	7	C	0	455	268	58	40	2.9
DIST	8	5	D	0	2631	880	255	121	21.0
DK17	10	11	C	1	631	138	57	19	3.7
DK27	9	9	C	1	200	46	20	10	2.8
DK48	15	17	C	1	672	144	42	22	10.6
EXEP	30	63	C	1	1944	1284	149	109	90.4
F51M	8	8	D	0	3064	396	255	76	19.3
GARY	15	11	C	0	2240	1114	214	107	17.0
IN0	15	11	C	0	1825	1114	135	107	23.7
IN1	16	17	C	0	2100	1970	110	104	37.2
IN2	19	10	C	0	1527	1448	137	135	24.2
IN3	35	29	C	0	852	771	75	74	16.1
IN4	32	20	C	0	3291	2543	234	212	69.9
IN5	24	14	C	0	752	738	62	62	8.2
IN6	33	23	C	0	552	552	54	54	6.0
IN7	27	10	C	0	563	427	84	54	7.4
JBP	36	57	C	0	1252	1043	166	122	122.8
MISG	56	23	C	0	255	247	75	69	7.1
MISH	94	43	C	0	255	238	91	82	6.7
MLP4	8	8	D	0	2478	897	225	127	33.2
OPA	17	69	C	0	2461	1108	342	82	86.8
RADD	8	5	D	0	768	415	120	75	5.8
RCKL	32	7	D	0	1238	657	96	32	19.7
RD53	5	3	D	0	197	175	31	31	1.2
RD73	7	3	D	0	1023	902	147	127	10.6
RISC	8	31	C	0	407	187	74	28	2.4
ROOT	8	5	D	0	2655	372	255	57	12.6
SQN	7	3	C	1	732	228	84	38	3.4
SQR6	6	12	D	0	637	290	63	49	11.8
TI	47	72	C	0	3171	2590	241	214	519.0
TIAL	14	8	D	0	5627	5159	640	579	682.8
VG2	25	8	C	0	914	924	110	110	10.7
WIM	4	7	C	1	91	50	10	9	1.4
X1DN	27	6	C	0	1090	1084	112	110	8.6
X2DN	82	56	C	0	578	567	112	105	32.3
X6DN	39	5	C	0	1400	817	121	81	11.1
X7DN	66	15	C	0	5642	4600	622	538	417.1
X9DN	27	7	C	0	1258	1268	120	120	10.3
Z4	7	4	C	0	1145	311	127	59	4.2
5XP1	7	10	D	0	1472	355	128	64	10.3
9SYM	9	1	D	0	4200	595	420	85	16.3

Table 6.1 ESPRESSO-IIAPL Results

ESPRESSO-II, and to describe the correlation between problem size and computing time. Another goal is to evaluate the algorithms of ESPRESSO-II in regard to their effectiveness and efficiency.

6.1 Analysis of Raw Data for ESPRESSO-IIAPL

In Table 6.1 we describe the 56 sample PLA's and the result of minimizing them with ESPRESSO-IIAPL. The PLA's are identified by alphanumeric names given in the first column. The next four columns specify the number of input and output variables, the "type" of the PLA (control logic (C) or data flow (D)), and whether or not a don't-care set is present. We then list the number of literals and terms in the incoming PLA (Init) and the final cover (Fin) obtained by ESPRESSO. (A literal is either a 0 or 1 in the input part, or a 4 in the output part.) The number of product terms in the final PLA determines the number of rows in its physical implementation, so we concentrate on this quantity as our measure of optimization. The final column gives the CPU time, in seconds, expended by ESPRESSO-II APL on an IBM/3081 K.

There is no direct relation between any one of the measures of problem size (inputs, outputs, literals, terms, complement size, etc.) and the computing time. However, there is some correlation as can be seen by scanning each column and noting that the larger numbers in any of the columns tend to cluster together. Of equal interest are the exceptions. For example, although TIAL (a 4-bit ALU) has only 14 inputs and 8 outputs, it has 581 product terms and requires 526 seconds to process. A similar statement can be made about ADD6, a six-bit adder. It would be of interest to see how accurately one can predict the computing time based on the initial PLA data (inputs, outputs, terms, literals). From the data given, it seems that some reasonable, possibly nonlinear, prediction formula could be constructed.

6.2 Analysis of Algorithms.

In this section we examine data regarding each of the major algorithms of ESPRESSO-II. For each PLA, we computed the percentage of total execution time expended by each major procedure. Table 6.2 describes the statistical behavior of CPU expenditure for our 56 sample problems. We now discuss the relative efficiency and cost effectiveness of each procedure.

Procedure	CPU Percentage Average	Standard Deviation	Minimum	Maximum
Complement	14	10	1	51
Expand	29	13	9	62
Irredundant Cover	12	6	2	28
Essential Primes	13	7	3	27
Reduce	8	6	0	26
Reduce-gasp	5	3	0	13
Expand-gasp	7	4	0	17
Lower Out	8	4	1	22
Raise in	2	2	0	9

Table 6.2

Complement and Essential Primes

COMPLEMENT and ESSENTIAL_PRIMES are each executed exactly once. Each could be omitted, but the cost of subsequent iterations would increase. In Table 6.3, some results specific to these two operations are reported. For each PLA, we report the number of essential primes it has and the percentage of execution time saved by detecting them. (The latter is computed by running ESPRESSO-II twice, once with the algorithm ESSENTIAL_PRIMES and once without.) We also list the number of terms and literals in the complement (recall no output sharing is allowed in our complement representation), and give the number of CPU seconds required to compute it.

We note from Table 6.2 that COMPLEMENT takes, on the average, about 14% of the total time. The complement is used in EXPAND, to quickly determine primes which cover a given cube, and to help guide which of the primes will be chosen to produce the best cover. No other procedure requires knowledge of the complement. Note, from Table 6.2, that EXPAND is the most expensive procedure; about 29% of the total time is spent in this routine. Without COMPLEMENT, this number would increase. Our initial experiments with ESPRESSO-I

PLA Name	Number Essen. Primes	% Time Saved	Complement Terms	Complement Lits.	Complement Time (sec)
ADD6	153	4.9	1093	11303	6.6
ADR4	35	5.6	86	564	2.5
ALU1	19	33.6	20	62	.1
ALU2	36	8.7	192	1478	.9
ALU3	27	25.2	235	1775	.8
APLA	0	-7.4	150	1392	1.6
BCA	144	39.3	4100	26131	16.6
BCB	137	41.5	3205	19930	12.5
BCC	119	37.8	3218	19052	13.7
BCD	100	34.7	2630	15405	9.0
BCO	37	11.1	590	5124	11.0
CHKN	86	48.2	323	2844	2.8
CO14	0	-7.1	92	652	.1
CPS	58	17.2	1979	8162	10.2
DC1	3	-9.4	28	101	.1
DC2	18	6.5	86	416	.5
DIST	23	30.0	181	1337	2.4
DK17	0	-16.0	108	1033	1.3
DK27	0	-106.0	106	1060	1.4
DK48	0	-14.7	232	3510	5.3
EXEP	82	37.7	3172	23741	15.7
F51M	13	8.3	79	438	4.3
GARY	60	11.8	308	2157	1.6
IN0	60	25.8	340	2323	2.4
IN1	54	16.3	1186	8461	5.8
IN2	85	22.0	337	2060	1.7
IN3	43	17.5	411	2017	1.9
IN4	118	22.8	499	3073	3.4
IN5	53	27.3	415	2456	1.7
IN6	41	33.3	336	1448	.9
IN7	35	12.5	113	606	.6
JBP	1	-41.7	537	1846	1.9
MISG	38	3.8	71	396	.3
MISH	73	42.7	86	265	.7
MLP4	12	21.2	201	1398	2.8
OPA	23	26.0	775	2918	3.0
RADD	35	1.8	121	853	.6
RCKL	6	-31.8	133	861	.3
RD53	21	14.3	36	192	.2
RD73	126	11.6	147	1031	.6
RISC	13	39.0	142	325	.4
ROOT	9	-6.5	89	549	2.0
SQN	23	9.3	53	304	.6
SQR6	3	18.0	76	359	1.0
TI	47	32.5	1266	6267	9.7
TIAL	220	.0	1021	9397	6.2
VG2	100	47.3	300	1876	1.1
WIM	3	-66.7	14	64	.3
X1DN	100	43.2	283	1805	1.1
X2DN	22	-21.9	196	660	1.3
X6DN	61	28.6	299	2446	2.0
X7DN	378	54.3	1074	8124	7.7
X9DN	110	45.0	329	1991	1.2
Z4	35	13.3	64	407	1.1
5XP1	8	13.2	80	399	2.3
9SYM	0	-22.5	72	684	1.5

Table 6.3 Essential Primes and Complement Data

suggest that the increase in computing time needed for the EXPAND operation more than offsets the 14% initial cost of COMPLEMENT.

The elimination of COMPLEMENT would also reduce the quality of the cover obtained after each expand step. In fact, the blocking matrix is essential to our expansion strategy. Unfortunately, we have not carried out precise measurements to quantify these effects. However, experimental results with POP as well as the comparison between PRESTO/POP and ESPRESSO-II indicate that this analysis is qualitatively correct.

We did experiment with improving the quality of the complement cover by using COMPWC. By checking for single cube containment during the merging process, this procedure produces a cover for the complement which is more nearly a prime cover. We found that a more minimal representation for the complement results in no improvement in the overall execution speed or the quality of the final results.

The ESSENTIAL_PRIMES procedure, on the other hand, appears to be cost-effective in general. Recall that once the essential primes have been identified, they are moved into the don't-care set during the main iteration; execution accelerates as a result. When the essential primes test is suppressed, average total execution time increased by 14%. Roughly speaking, the 13% investment in essential prime identification typically pays double for itself.

Of course, some PLA's have few if any essential primes. For instance, JBP has only 1 essential prime but a minimal cover of 122 cubes, and 9SYM has no essential primes. Note that DK17, DK27, and DK48 also have no essential primes. These are PLA's which have a don't-care set representing a multi-valued variable as discussed in Chapter 5. In these three cases one can show, a priori, that no essential primes exist, so it would have been possible to deactivate the essential primes calculation and save computing time. In general, it is not possible to predict the number of essential primes, so the best strategy seems to be to always try to find them.

Iterated Operations

For those procedures which are performed more than once, we give in Table 6.4 a breakdown of execution time according to the iteration on which the step is performed. Adjacent to the average percentage of CPU time expended, we give the number of sample cases (out of 56 possible) in which that iteration is actually reached. If processing terminates before a given iteration, we average in zero for the time expenditure; thus the percentages in columns 4, 5 and 6 are small because most problems terminate in 3 or fewer cycles. The percentages in each row sum to the total percentage given in Table 6.2.

CPU Percentage/Number of Occurrences

	Iteration Count					
Procedure	1	2	3	4	5	6
Expand	21.0/56	5.1/54	1.6/21	0.8/11	0.2/5	0.0/2
Irred-Cover	9.4/56	1.7/35	0.4/11	0.3/8	0.2/5	0.0/2
Reduce	4.9/54	1.7/21	0.8/11	0.4/5	0.1/2	
Reduce-gasp	4.5/54	0.5/8	0.1/1			
Expand-gasp	5.8/54	0.7/8	0.1/1			

Table 6.4 Average Percent of Total Time Spent During Iterations

Note that each procedure is most expensive during the first iteration, particularly EXPAND. This is primarily because the cover is at its largest initially. After the first expansion step, the cover is typically reduced to 80% of its original size; even more dramatic is the elimination of essential primes following irredundant cover, after which only 34% of the original cubes remain, on average. Hence, subsequent iterations are faced with a substantially smaller problem.

We remark that the execution time of EXPAND is correlated with the size of the complement, since this determines the size of the blocking matrix. There are six PLA's among the 56 sample cases for which the size of the complement exceeds 1900 terms; for these six, EXPAND consumes an average of 60% of the total CPU time, more than twice the average time expenditure for the remaining cases, all of which have complements of fewer than 1300 terms.

Evaluation of the cost-effectiveness of the procedures IRREDUNDANT_COVER, REDUCE, LASTGASP and MAKESPARSE was carried out by removing these steps from the ESPRESSO-II procedure and observing the change in execution time and quality of results (number of terms in the minimized cover). This data is summarized in Table 6.5; we list, for each procedure, the percentage by which it reduces the cover size, and the percentage of CPU time expended in doing so.

Procedure	Cover Reduction	Time Increase
IRREDUNDANT_COVER	1 %	−0.2 %
REDUCE	5.1 %	36.6 %
LASTGASP	0.5 %	11.98 %
MAKESPARSE	7.7 % (literals)	10.9 %

Table 6.5 Average Effectiveness of Different Procedures

To interpret this table, it is important to understand how ESPRESSO-II functions when a given procedure is removed. When IRREDUNDANT_COVER is skipped, REDUCE still produces a minimal cover, because redundant cubes are reduced until empty. On the other hand, when REDUCE is skipped ESPRESSO-II does not iterate; the incoming data to EXPAND on the second pass is already a prime irredundant cover. Thus the 5.1% figure for REDUCE partly measures the effectiveness of iterative improvement. Also note that MAKESPARSE does not reduce the number of terms in the cover, only the number of literals; therefore we have measured its effectiveness in terms of literal reduction. Finally, we remark that although LASTGASP produces a relatively small reduction in cover size, its purpose is to provide confidence in the quality of the final result; the fact that LASTGASP finds an additional improvement in only 8 of the 56 test cases is to some extent a consequence of the effectiveness of the other algorithms.

Since the data in Table 6.5 represents average behavior, it should be taken with a grain of salt; performance on a particular problem may depend on its size and other properties. For example, on the largest test problem, ESPRESSO-II requires over 1300 CPU seconds without IRREDUNDANT_COVER; this is reduced to 721 seconds when the procedure is included.

6.3 Optimality of ESPRESSO-II Results

For 32 of the 56 test PLA's, R. Rudell has determined the exact or approximate size of the absolute minimum cover. Table 6.6 summarizes this data. We give for each of these 32 PLA's the total number of primes, the number of essential primes, and the size of three different covers. Under MIN we list the size of the **minimum** cover; when this is not known exactly, a range of possible values is given. Under ESP-APL we list the size of the cover obtained by ESPRESSO-IIAPL, and under Q-M we give the size of the cover obtained by applying IRREDUNDANT_COVER to the set of **all** primes. This "modified Quine-McClusky" algorithm is also used to compute a lower bound for the size of the minimum cover.

In 5 of the 32 problems, the minimum is only known to within 1 to 4 cubes. In 21 of the 32 cases ESPRESSO-II actually achieves the minimum cover while Q-M achieves the minimum for 24 of these problems. In 4 cases ESPRESSO-IIAPL achieves a superior solution while in 5 cases Q-M is better. In total, for these 32 problems, ESPRESSO-IIAPL obtained 5 fewer cubes than Q-M did. Of course ESPRESSO-II is much faster than Q-M, and indeed in the remaining 24 test cases it is infeasible to generate the set of all prime implicants.

In Table 6.7 we compare the C-version of ESPRESSO-II with the modified Quine-McClusky algorithm Q-M, which is also implemented in C. The last two columns give a breakdown of the execution time for Q-M according to the two steps involved, the generation of all primes and the extraction of an irredundant cover. On the larger problems, the total CPU time expended by Q-M is more than 10 to 100 times that required by ESPRESSO-IIC. Surprisingly, the two algorithms generate comparable results for the 32 sample problems. In conclusion,

Optimality of ESPRESSO-II Results

PLA	In	Out	Primes	Essen	Cover Size MIN	ESP-APL	Q-M
ADR4	8	5	397	35	75	75	75
ALU1	12	8	780	19	19	19	19
ALU2	10	8	434	35	68	68	68
ALU3	10	8	540	27	64	65	64
APLA	10	12	201	0	25	25	25
CHKN	29	7	671	86	140	140	140
CO14	14	1	14	14	14	14	14
CPS	24	109	2487	57	157-159	160	163
DC1	4	7	22	3	9	9	9
DC2	8	7	173	18	39	40	39
DIST	8	5	401	23	120	121	121
DK17	10	11	111	0	18	19	18
DK27	9	9	82	0	10	10	10
DK48	15	17	157	0	21	22	21
F51M	8	8	561	13	76	76	76
IN0	15	11	706	60	107	107	107
IN1	16	17	928	54	104	104	107
MLP4	8	8	606	12	119-123	127	128
OPA	17	69	477	22	76-77	82	82
RADD	8	5	397	35	75	75	75
RCKL	32	7	302	6	32	32	32
RD53	5	3	51	21	31	31	31
RD73	7	3	211	106	127	127	127
RISC	8	31	46	22	28	28	28
ROOT	8	5	152	9	57	57	57
SQN	7	3	75	23	38	38	38
SQR6	6	12	205	3	44-46	49	48
VG2	25	8	1188	100	110	110	110
WIM	4	7	25	3	9	9	9
Z4	7	4	167	25	59	59	59
5XP1	7	10	390	8	61-63	64	64
9SYM	9	1	1680	0	84	85	88

Table 6.6

ESPRESSO-II seems to perform just as well as the much more expensive and theoretically more precise modified Quine-McClusky algorithm.

Thus we have verified that ESPRESSO-II produces optimal or near optimal results in those cases where such a verification is feasible. The virtue of ESPRESSO-II is that it produces results at an efficient rate (and presumably of similar quality) when faced with problems intractable by more exact methods.

We now discuss the method by which the size of the absolute minimum cover is estimated. First the set of all primes is generated,

Comparison of C Versions
of ESPRESSO and Modified Quine-McClusky

PLA	Cover Size		CPU Time		Breakdown	
	ESP-C	Q-M	ESP-C	Q-M	= PRIME	+ IRRED
ADR4	75	75	2.0	6.6	2.9	3.7
ALU1	19	19	.1	42.9	32.9	10.0
ALU2	68	68	2.8	28.3	11.5	16.8
ALU3	65	64	4.8	42.9	18.3	24.6
APLA	25	25	.9	20.5	14.2	6.3
CHKN	141	140	13.0	591.3	566.6	24.7
CO14	14	14	.1	.2	.1	.1
CPS	161	163	107.4	1829.7	568.5	1261.2
DC1	9	9	.1	.2	.1	.1
DC2	39	39	.7	2.0	1.0	1.0
DIST	120	121	11.6	25.2	17.2	8.0
DK17	18	18	.7	7.9	6.2	1.7
DK27	10	10	.6	6.3	5.4	.9
DK48	22	21	3.7	89.5	85.3	4.2
F51M	76	76	5.1	20.8	8.8	12.0
INO	107	107	4.5	104.7	76.9	27.8
IN1	104	107	6.8	554.7	521.5	33.2
MLP4	127	128	9.0	47.3	23.7	23.6
OPA	80	82	19.6	46.2	21.7	24.5
RADD	75	75	1.3	6.4	2.4	4.0
RCKL	32	32	5.3	24.2	3.3	20.9
RD53	31	31	.2	.3	.2	.1
RD73	127	127	2.5	3.2	2.4	.8
RISC	28	28	.6	.4	.3	.1
ROOT	57	57	3.4	4.0	2.3	1.7
SQN	38	38	.6	.8	.5	.3
SQR6	50	48	2.1	4.5	1.8	2.7
VG2	110	110	3.8	598.2	572.2	26.0
WIM	9	9	.2	.2	.1	.1
Z4	59	59	1.0	2.0	1.2	.8
5XP1	63	64	3.5	13.6	4.8	8.8
9SYM	87	85	4.6	376.9	189.2	187.7

Table 6.7

using a new method proposed by R. Rudell. Then **IRREDUNDANT_COVER** is invoked, to generate the matrix $\overline{\mathbb{B}}$ which gives the condition that a subset of the set of all primes is itself a cover. In principle, one can now apply a branch-and-bound algorithm to find the *minimum* cover; however, this is practical only for very small problems. Instead, we apply the MINUCOV algorithm, which uses an independent set heuristic. This heuristic gives as a by-product a lower bound on the size of the minimum cover.

We compare this lower bound with the size of actual covers generated by the APL and C versions of ESPRESSO, as well as the Q-M algorithm. The true minimum lies somewhere between the lower bound given by MINUCOV and the size of the covers obtained by these three algorithms. In all but five of the 32 cases, at least one of the three algorithms actually achieves the minimum covering. In the five exceptional cases, the range of possible values for the minimum was narrowed by experimenting with the algorithms and heuristics until a better covering was obtained.

Chapter 7

COMPARISONS AND CONCLUSIONS

In this chapter we recapitulate the main features of ESPRESSO-II and compare it to other minimization algorithms. We compare its strategy with that of MINI, POP and PRESTO, and discuss some modifications introduced in the C-language version of ESPRESSO-II. We then present empirical results obtained by running these programs on the 56 test cases introduced in the preceding chapter. Finally, we discuss other applications of logic minimization and directions for further research.

7.1. Qualitative Evaluation of Algorithms of ESPRESSO-II

The algorithms of ESPRESSO-II form a logic minimization tool which achieves both robust performance and quality results. Iterative improvement produces well-minimized covers with high confidence, while the unate paradigm together with special-case handling ensures reasonably efficient execution for a broad range of incoming problems. We summarize below those features of the ESPRESSO-II algorithm which contribute to this level of performance.

Speed of execution is, to a large part, due to the behavior of algorithms based upon the unate paradigm. Typical PLA's, particularly control logic, produce shallow recursion trees terminating quickly at unate leaves. This benefit is realized in the basic Boolean manipulation algorithms: complementation, and, especially, in the repeatedly executed tautology computation. These algorithms also employ several heuristics to trim the recursion tree and balance it judiciously.

Robust performance across a wide spectrum of problems depends on the handling of uncommon but occasionally encountered special cases. For example, Vanilla tautology will run rather slowly on a problem which has a natural block structure; however, the tautology algorithm used in ESPRESSO-II will detect if there is a reasonable possibility that such a structure exists, and invoke the COMPONENT__REDUCTION procedure. Block structures also lead to small functions with exponentially large complements, such as the Achille's heel function (described in Section 4.9). When a large complement is encountered, the "output splitting" strategy virtually guarantees reduction of the complement size to manageable proportions. Some such dynamic partitioning algorithm is essential to ensure reliable, intervention-free executability. In practical VLSI applications, we have encountered one problem with 152 inputs and 382 outputs, and another with 3236 product terms; even in these cases, processing proceeds routinely and does not require unreasonable resources when output splitting is employed. Such reliability is required if a logic minimizer is to function as part of a Silicon Compiler.

The quality of the minimization obtained by ESPRESSO-II is the result of iterative improvement through EXPAND, IRREDUNDANT__COVER, REDUCE and LAST__GASP. The novel blocking and covering matrices employed in EXPAND allow a large set of primes to be explored as potential lifts of a given reduced cube. The LAST__GASP algorithm assures that our final result is weakly optimal, in the sense that no better solution exists near to it. More precisely, no single prime can be added to the final result in such a way that two primes can then be discarded.

The expansion/reduction cycle permits exploration of a large space of possible covers; considering in addition the weak optimality just mentioned, we generally have confidence that our final result is well-minimized. This conclusion is supported by the results of Section 6.3, which show that ESPRESSO-II obtains results close to the absolute minimum at least in those cases where the minimum has been determined.

7.2 Comparison with ESPRESSO-IIC

For the main algorithms, ESPRESSO-IIC (the C language version) follows exactly the procedures outlined in this book. Its differences lie in the heuristics used to break ties, to order the cubes in the cover, and in some improvements in MINUCOV. Table 7.1 shows the results obtained by ESPRESSO-IIC for the 56 PLA's. Regarding the number of terms in the minimized cover, ESPRESSO-IIC and ESPRESSO-IIAPL obtain the same result in 43 of the 56 cases; in 7 cases the C version obtains the better solution, while in 6 cases the APL version wins out. Overall, the quality of results obtained are nearly identical. The total cover for all 56 PLA's obtained by ESPRESSO-IIC was 6001 cubes and for ESPRESSO-IIAPL it was 5998 cubes. The main differences between ESPRESSO-IIC and ESPRESSO-IIAPL will be summarized below. Future versions of both programs may incorporate these or other improvements.

Ordering for Reduce

Early experiments indicate that using the following ordering before REDUCE may be effective.

1. Order the cubes according to decreasing size where the size of a cube is determined by the sum of the number of 2's in the input and the number of 4's in the output.
2. Find the first cube which can be reduced.
3. Reorder the cubes in decreasing distance from the cube found in 2.

Improvements in MINUCOV

Recall that MINUCOV employs a maximal clique algorithm to determine a large independent set of rows in the matrix to be covered. The algorithm used in MAXCLIQ to find a maximal clique in a graph G is as follows:

0. Let Q be empty.
1. Find the set S of nodes of G with maximum adjacency.
2. Find the largest clique in G restricted to S.
3. Choose k from this clique and put k in Q.
4. From G, eliminate node k and all nodes not adjacent to k.
5. Repeat 1, 2, 3 and 4 until G is empty.

This differs from the algorithm used in ESPRESSO-IIAPL in that, instead of 2 and 3, the first node of maximum adjacency is chosen.

The second improvement in MINUCOV comes from the way a column from each independent set is chosen. In ESPRESSO-IIAPL, the independent sets were ordered in decreasing size and the column with maximum column count from each set is chosen in succession. In ESPRESSO-IIC, of all the columns associated with all the independent sets, the one with the largest weight is chosen. A column c has its weight determined by the inner product of c and w, $\Sigma c_i w_i$, where w_i is the inverse of the ith row sum of the matrix A to be covered, i.e.

$$w_i = \left(\sum_{j=1}^{n} A_{ij} \right)^{-1} .$$

After a column is chosen, the rows covered by this column are eliminated from A, the associated independent set is eliminated, and the procedure is repeated until no independent set remains.

These two changes represent a distinct improvement in the minimum covering algorithm. This becomes evident when used in the "modified Q-M" algorithm. Its effect on ESPRESSO-II, however, is not as pronounced.

Minterm Expansion

ESPRESSO-IIC uses a heuristic which was the source of the use of the UNWRAP procedure described in Section 4.0. Most of the benefit of UNWRAP comes from examples where the input part has no 2's, i.e. the input parts are all minterms. In ESPRESSO-IIC, UNWRAP is employed only if the input part of every cube is a minterm. In this case, each cube of the cover is split into several cubes such that each resulting cube has only a single 4. (By splitting the output part, we have made the entire cube a minterm.) This increases the size of the cover, but provides more degrees of freedom for the initial expand. In particular, the Maximal Feasible covering set heuristic of procedure EXPAND (cf., Section 4.3.5) will take each minterm and expand it using the greedy heuristic of covering as many other minterms as possible. Results from the 9 examples which are affected by this heuristic (ADR4, DIST, F51M, MLP4, ROOT, SQR6, WIM, Z4, 5XP1) show that an equivalent or better solution is achieved in each case. In particular, substantially better results are observed for the examples MLP4, SQR6 and 5XP1. By not using UNWRAP on the other kinds of PLA's, equivalent results are obtained but in less time.

7.3 Comparison of ESPRESSO-II with Other Programs

At the time the development of the ESPRESSO-II algorithms was started, the most successful practical PLA logic minimizers were the various FORTRAN and APL implementations of MINI. PRESTO, [BRO 81] had just been announced. Since that time both ESPRESSO-II and POP [DEM 84] have been developed. In this section we will first discuss the minimization strategies employed by these programs, then compare their results with those of ESPRESSO-II on the 56 test PLA's. (We have no test data for PRESTO, but POP, a derivative of PRESTO, gives representative results.)

PRESTO processes the implicants in the order given by the user. The cubes are first maximally expanded in their input part. To check if an input can be raised, PRESTO checks if the expanded cube is covered by the other cubes of the function. The covering check is performed by a tautology algorithm reminiscent of the technique developed by Morreale [MOR 70]. Since the complement of the logic function is not

Name	In	Out	Terms Ess C	Terms Ess APL	Terms Min APL	Terms POP C	Literals Ess C	Literals Ess APL	Literals Mini APL	Literals POP C	Time(sec.) Ess-C 3081	Time(sec.) Ess-A 3081	Time(sec.) Mini-A 3081	Time(sec.) POP-C VAX
ADD6	12	7	355	355	355	355	2551	2551	2551	2551	28.4	118.2	1051.0	946.3
ADR4	8	5	75	75	75	75	415	415	415	415	3.0	9.3	52.6	114.9
ALU1	12	8	19	19	19	19	60	60	60	60	.1	.6	8.2	1.2
ALU2	10	8	68	68	69	74	347	347	353	357	2.8	12.0	178.5	19.2
ALU3	10	8	65	65	66	68	351	351	349	352	4.8	13.4	125.9	19.6
APLA	10	12	25	25	25	57	222	222	220	293	.9	7.7	38.5	40.4
BCA	26	46	180	180	180	182	3265	3263	3298	3315	51.2	230.9	417.9	220.3
BCB	26	39	155	155	156	156	2766	2765	2810	2788	28.3	125.2	318.1	151.4
BCC	26	45	137	138	138	138	2529	2542	2552	2549	31.1	121.2	275.6	184.6
BCD	26	38	117	117	117	118	2022	2024	2030	2036	18.2	77.8	207.3	158.3
BCO	26	11	179	179	179	183	2138	2084	2117	2165	24.9	101.8	551.3	383.4
CHKN	29	7	141	140	141	145	1748	1739	1752	1799	13.0	50.9	1008.5	225.7
CO14	14	1	14	14	14	14	210	210	210	210	.1	.4	13.9	3.1
CPS	24	109	161	160	165	400	2917	2893	2896	5205	107.4	723.9	1378.4	559.1
DC1	4	7	9	9	9	14	54	54	58	64	.1	.9	2.3	15.0
DC2	8	7	39	40	39	42	265	268	259	271	.7	2.9	24.4	8.0
DIST	8	5	120	121	123	128	868	880	880	940	11.6	21.0	167.0	97.1
DK17	10	11	18	19	18	37	135	138	141	162	.7	3.7	16.6	30.1
DK27	9	9	10	10	10	16	46	46	51	55	.6	2.8	5.7	12.5
DK48	15	17	22	22	22	32	132	144	129	114	3.7	10.6	37.3	168.2
EXEP	30	63	109	109	108	112	1284	1284	1278	1303	25.2	90.4	356.2	134.4
F51M	8	8	76	76	76	76	397	396	403	399	5.1	19.3	89.1	141.2
GARY	15	11	107	107	108	107	1115	1114	1129	1120	4.0	17.0	227.4	66.0
INO	15	11	107	107	107	108	1115	1114	1125	1125	4.5	23.7	218.4	63.4
IN1	16	17	104	104	105	105	1970	1970	1951	1982	6.8	37.2	263.5	80.9
IN2	19	10	137	135	135	135	1512	1448	1439	1466	4.3	24.2	638.5	61.9
IN3	35	29	74	74	74	75	771	771	777	767	4.1	16.1	175.2	37.7
IN4	32	20	212	212	211	212	2545	2543	2577	2544	19.5	69.9	1909.3	258.8
IN5	24	14	62	62	62	62	741	738	738	738	1.9	8.2	184.8	25.3
IN6	33	23	54	54	54	54	552	552	549	547	1.7	6.0	196.6	16.2
IN7	27	10	54	54	55	55	427	427	422	417	1.9	7.4	104.4	43.9
JBP	36	57	122	122	123	166	1053	1043	1028	1252	22.2	122.8	1038.2	59.2
MISG	56	23	69	69	69	69	247	247	247	247	2.8	7.1	8819.1	62.5
MISH	94	43	82	82	82	82	238	238	238	238	5.0	6.7	837.3	30.4
MLP4	8	8	127	127	126	130	908	897	924	923	9.0	33.2	269.3	121.0
OPA	17	69	80	82	80	184	1120	1108	1127	1463	19.6	86.8	252.1	106.4
RADD	8	5	75	75	75	75	415	415	415	415	1.3	5.8	47.4	16.9
RCKL	32	7	32	32	32	96	657	657	657	1238	5.3	19.7	248.9	116.1
RD53	5	3	31	31	31	31	175	175	177	175	.2	1.2	7.1	3.9
RD73	7	3	127	127	127	147	903	902	905	1023	2.5	10.6	88.1	38.2
RISC	8	31	28	28	28	30	187	187	189	184	.6	2.4	13.7	8.2
ROOT	8	5	57	57	58	58	353	372	371	400	3.4	12.6	28.8	53.4
SQN	7	3	38	38	38	39	230	228	231	235	.6	3.4	15.5	14.0
SQR6	6	12	50	49	49	53	273	290	259	292	2.1	11.8	56.3	20.5
TI	47	72	214	214	222	222	2587	2590	2610	2584	154.3	519.0	7988.5	253.8
TIAL	14	8	583	579	578	640	5181	5159	5112	5627	146.7	682.8	8306.9	748.9
VG2	25	8	110	110	110	110	924	924	934	914	3.8	10.7	361.5	50.9
WIM	4	7	9	9	9	10	47	50	50	64	.2	1.4	2.2	3.1
X1DN	27	6	110	110	110	110	1084	1084	1074	1074	2.7	8.6	398.2	60.6
X2DN	82	56	104	105	104	110	564	567	564	569	48.4	32.3	1141.7	44.9
X6DN	39	5	81	81	81	87	817	817	822	862	3.5	11.1	226.8	75.3
X7DN	66	15	538	538	548	622	4600	4600	4691	5356	135.8	417.1	3245.0	912.5
X9DN	27	7	120	120	120	120	1268	1268	1264	1258	3.2	10.3	482.7	63.8
Z4	7	4	59	59	59	59	311	311	311	311	1.0	4.2	29.0	35.7
5XP1	7	10	63	64	73	76	357	355	523	593	3.5	10.3	68.5	42.5
9SYM	9	1	87	85	85	148	609	595	595	1036	4.6	16.3	226.3	212.3

Table 7.1 Comparison of PLA minimization Programs

needed, PRESTO is suitable for minimization of logic functions with very large complements.

Once the inputs are maximally raised, the output part of the implicant is maximally reduced. This step is performed by lowering the outputs one at a time and checking if the cube so removed is covered by the remaining cubes of the function. The covering check is identical to the one performed during the expansion phase. Hence PRESTO is almost entirely based on tautology. This output reduction phase guarantees that the cover is irredundant at the end of the first pass through all the implicants. However, since the outputs are never raised, the cubes in the cover need not be prime. The entire procedure is iterated until no variation is observed. This reduction sequence does leave the cover matrix maximally sparse at the end. The success of PRESTO is limited by its lack of several important components: an efficient tautology algorithm, a good heuristic to direct its expansion steps, and a reduce or reshape procedure to permit escape from local minima.

POP grew out of work done at Berkeley on PRESTO. It provides a better cube ordering strategy and more efficiently implemented algorithms. During the development of ESPRESSO-II, it became clear that the expansion step was more efficiently implemented in most cases by using the complement. P. Szimanyi added a complementation option to POP using an algorithm based on the unate paradigm explained in Chapter 3. This mode of operation generally yields faster execution of the program.

MINI, developed at IBM in 1974, invests more time than POP in an effort to obtain a well-minimized cover. It raises the outputs as well as the inputs to prime and performs a full cube reduction during each iteration. MINI also uses a "reshape" strategy, based on consensus, to move away from local minima. Finally, MINI allows multi-valued inputs, and therefore has the power to perform true 3-level minimization. (As discussed in Chapter 5, this capability is also available, in principle, to any two-level minimizer which allows a don't-care set.)

ESPRESSO-II incorporates most of the features of MINI, but supplies more efficient and sophisticated algorithms for the principle operations. In addition to improvements resulting from the unate paradigm, novel features of ESPRESSO-II include the blocking and covering matrices employed during expansion, and the $\overline{\mathbb{B}}$ matrix (Section 4.5), used during irredundant cover. Using the blocking and covering matrices, we can guarantee that any single-cube expansion which reduces the size of the cover will be discovered; MINI gives no such assurance. The $\overline{\mathbb{B}}$ matrix translates the problem of choosing the minimum subcover of a prime cover into a standard "column-covering" problem, which can be solved exactly by modified unate complementation. In the interest of efficiency we employ instead a heuristic column-covering algorithm, which we find achieves the minimum in most cases.

The price of a minimization usually increases with the degree of confidence required in the minimality of the result. Sometimes simpler programs achieve the same result more quickly, or even, rarely, find a better result. We believe, however, that over a full range of test problems, ESPRESSO-II provides a high confidence result most efficiently, and represents a unique capability at the upper end of the spectrum.

Table 7.1 gives the results obtained by ESPRESSO-IIAPL, ESPRESSO-IIC, MINI-APL and POP for the 56 test PLA's introduced in Chapter 6. The first three results were obtained on the same machine, an IBM-3081K, so the times given can be compared directly. POP is coded in C and was run on a VAX 11-780; for comparison, we note that we experienced a factor of 12 speedup when ESPRESSO-IIC was transported from the VAX 11-780 to the IBM 3081K.

Table 7.2 gives the column totals for each of the minimization programs:

	Totals		
	Terms	Literals	Time
ESPRESSO-IIAPL	5998	60432	4002.2
ESPRESSO-IIC	6001	60578	992.9
MINI-APL	6032	60837	44441.5
POP-C (VAX)	6828	66442	7443.1

Table 7.2

In terms of the quality of the final cover, ESPRESSO-IIAPL and ESPRESSO-IIC are roughly equivalent, while they both give slightly better results than MINI. A more distinct improvement is in the efficiency of the algorithms where ESPRESSO-IIAPL is a factor of 11 faster than MINI-APL. This ratio is quite variable. Across the 56 PLA's, the median of this ratio was 8.16 and the average was 36.1. POP in many cases gives a good result but it sometimes gets stuck at a very bad answer and hence is somewhat unreliable. On the 56 PLA's, POP averages 20% more cubes than ESPRESSO-II, but the median is 2.2% more cubes, indicating that when POP gets stuck, the result can be quite unsatisfactory. Using the rough estimate of 12 between the 3081K and the VAX, we compute that the ratio of total POP time to total ESPRESSO-IIC time (projected to the VAX) is .63. However, on the 56 PLA's, the median for this ratio is 1.12 and the average is 1.57. Thus on most of the PLA's ESPRESSO-IIC (VAX) is faster. These statistics can be understood by noting that POP appears to be faster on 21 of the PLA's, but on these it gets a total of 574 more cubes than the 3157 which ESPRESSO-IIC produces for these 21 PLA's.

Scanning Table 7.1, one may observe that on the whole, the quality of the minimized results obtained by MINI and ESPRESSO-II are remarkably similar. Both programs give about a 50% reduction in total literal count and a 40% reduction in product term count. As discussed in Section 6.3, this is due to the fact that in most cases, both MINI and ESPRESSO-II give nearly optimum results.

Table 7.3 gives a (somewhat incomplete) comparison with MINI-II, an APL version of MINI which was improved by T. Sasao, [SAS 83a], by adding a fast complementer (similar in some aspects to our unate complementer) and a fast cube reduction algorithm. The modified program runs faster than the original MINI and yields somewhat better results. The MINI-II test runs shown were obtained on a slower running IBM 3081 than the ESPRESSO-II test runs; to make a comparison, the ESPRESSO-II times should be multiplied by a factor of 2. Allowing for this difference, ESPRESSO-II executes about five times faster than MINI-IIAPL, and obtains comparable results.

The 56 test problems selected include a wide range of sizes and types of PLA's, and our test runs probably provide a good indication of the practical performance of the minimization programs. We feel there is sufficient evidence to support the claim that ESPRESSO-II represents a substantial advance in practical PLA minimization.

The implementation of ESPRESSO-II in the C language, developed by R. Rudell, which exploits features of the C language to provide a machine independent version runs approximately four times faster than the APL version in the same machine. ESPRESSO-IIC has been run on the DEC VAX 11-780, IBM 3081, the SUN and APOLLO 68000-based workstations, as well as on the IBM PC personal computer with a minimum of conversion effort. Also, there is a partial APL version of ESPRESSO-II, developed at Boulder, implemented on the IBM PC and XT personal computers.

PLA	INITIAL SIZE	MINI-II SIZE	MINI-II TIME*	ESP-IIAPL SIZE	ESP-IIAPL TIME*
DC2	58	39	21.4	40	2.9
IN1	110	105	92.4	104	37.2
IN2	137	135	246.	135	24.2
IN4	234	212	518.	212	69.9
IN5	62	62	152.	62	8.2
IN6	954	54	148.	54	6.0
IN7	74	54	107.	54	7.4
RISC	84	28	17.0	28	2.4
9SYM	420	85	410.	85	15.3
	1233	774	1724.7	774	174.5

* Timing carried out on different computers (see text).

Table 7.3 Comparison with APL MINI-II

7.4 Other Applications of Logic Minimization

We have emphasized throughout this book the application of logic minimization to PLA design. This is appropriate, since the impact of 2-level minimization is most immediate in the case of PLA's. However, it is also important to emphasize the role of logic minimization in other design methodologies. Minimization is important in designing the cells (or "books") of all cell libraries, as well as in the task of reducing Boolean expressions prior to decomposing them into cells. This becomes especially important in the context of high level logic languages, since as they approach true "silicon compiler" status, the designer is increasingly divorced from the internals of the design process, and left closer and closer to the top level of the design. The silicon compiler itself must become an optimizing compiler and automatically invoke appropriate logic minimization procedures. One ramification is the need for the high level language to facilitate the specification or extraction of the "don't-care" set accompanying a logic function. The don't-care set is often vital to the success of minimization; at present, we know of no logic specification language which produces a don't-care set for use by subsequent logic synthesis.

One obstruction to the use of minimization in popular design methodologies such as gate arrays and standard cells is the fact that the target cells are not uniform, e.g. not all "NOR" gates as they are in PLA's. Often more complicated cells, individually of greater logical power, are often more efficient. Further, the initial logic specified for gate-array or standard cell physical design programs is almost always multi-level rather than two-level. To date, these facts have been sufficient to obstruct the full exploitation of logic minimization in these technologies. However, we believe that ultimately a class of logic decomposition and minimization techniques will be produced which will become the basis of sophisticated "logic editors" for carrying out

1) optimized multi-level decompositions of specified logic;

2) "functional mapping" from one target technology to another.

One system for carrying out the above operations has already been reported [BRA 84]. In this system, ESPRESSO-II is essentially embedded as a local minimizer applied to the cover of cubes at each node of a multi-level "Boolean Network". A don't-care set is derived for the Boolean function f at such a node in a manner suggested by the treatment of Chapter 5. If f depends on input G and G is the output of an intermediate function g then $\mathcal{D}_g = g\overline{G} + \overline{g}G$ is don't-care for f. Repeating this for each non-primary input of f yields a don't-care set for f which can be exploited (often to significant effect) when applying

ESPRESSO-II to f. This step can be repeated, if desired, for the non-primary inputs of g, and, in fact for all intermediate variables in the cone of dependence of the given function f. This step is often too expensive to carry out completely, but it demonstrates the potential for exploiting the global structure of the problem. This application of ESPRESSO-II stands out in contrast to local transformation logic synthesis programs such as the LSS program described by Darringer, et al, [DAR 81]. A set of local transformations is used to simplify multi-level logic. All possible transformations in a given set are tested for applicability to each gate in the network. Optimization in the sense of local improvement is achieved whenever such a rule-action pair "fires". Each such local transformation can be viewed as a classical 2-level minimization step (or a sequence of such steps) applied to a piece of the network. However, there is a fundamental distinction between the local transformation approach and ESPRESSO-II, namely, general logic minimization applied to multi-level logic can account for global structure.

Another non-PLA methodology where optimization is already being exploited, is BDD synthesis (*B*inary *D*ecision *D*iagram, Akers [AKE 79], Matos [MAT 82], and Bartlett, et al [BAR 84]). BDD synthesis proceeds by recursive Shannon factorization until the leaves of the binary recursion tree are either empty (connect to ground) or tautological (connect to power). The Shannon factorization operation is implemented as a multiplexer macro-cell. Minimization techniques are applied to each cofactor at each node of the tree to minimize the depth of the tree.

7.5. Directions for Future Research

It is our hope that the ideas and algorithms described in this book represent a plateau for the development of two-level binary logic minimization. Many of the difficulties apparent to our precursors now have practical (and even, at times, elegant) solutions. Despite this progress, each algorithm and heuristic performs better in some cases than in others; the minimization problem is in principle difficult, and future developments will undoubtedly exploit fundamentally new ideas.

Here we briefly describe three directions in which future research might be pursued.

ESPRESSO-II Minimization with Multivalue Inputs

Multi-valued logic (e.g., [HON 74], [FLE 75], [SAS 78], [SAS 81], [SAS 82], [SAS 83]) is a substantial field in its own right. One of the major outstanding problems is the extension, of the ESPRESSO-II algorithms to true multi-valued inputs. For example, one can immediately query how to extend the definition (cf. Chapter 3) of a "unate cover" to the multivalued case. As discussed in Chapter 5, ESPRESSO-II can, presently, be applied to multi-valued input problems, which is important because often PLAs are mapped into a multi-valued input format by high level languages, e.g., IDL, [MAI 82], [MAI 82], [MAI 83], [WOO 83].

Although MINI already achieves true 3-level minimization, the prospect of n-level minimization, or the (nearly equivalent) problem of minimizing a chain of linked PLA's, is an area of active research. As computer aided design becomes more powerful, the rectangular regularity of a two-stage PLA becomes less important and an optimal logic design may take advantage of the freedom of a many-level cascade.

Completeness of a Set of Transformations

Like most minimizers, ESPRESSO-II applies a certain repertoire of operations to incoming data in an attempt to transform it to an equivalent optimal form. The order in which these operations are performed, and the cubes to which they are applied, are selected by various heuristics. Certainly a *necessary* condition for the effectiveness of the minimizer is that the set of available transformations is sufficiently rich to allow any incoming data to be converted to its minimum form. For example, one might ask if there is **some** sequence of cube expansions, choices of irredundant cover, and cube reductions, that eventually leads to the minimum prime cover. If this is not the case, then no matter how good the heuristics are, there are certain problems on which ESPRESSO-II would be doomed to fall short of the minimum.

At present, we know of no practical minimizer whose transformation set has been shown to be "sufficiently rich" in the sense above.

Alternate Representations of Logic Functions

We expect that the success of practical logic minimization, in view of the host of NP-complete problems encountered, is due at least in part to the special properties of logic functions encountered in real applications. Such functions are always much simpler than a general or random function of the same number of variables. Clearly it is desirable that a simple function should have a simple data representation. Our PLA-style data structure does not always have this property. For example, difficulties are sometimes encountered because a relatively small function has a large complement, while clearly both have the same order of logical complexity. Using a sequence of exclusive OR's, it is easy to describe a parity function on n variables. However, in our data representation such a function has 2^{n-1} terms.

Of course it is easy to create a compact representation of such functions by including complement and exclusive-OR in the primitive vocabulary of the data structure. What becomes much more difficult is the construction of efficient algorithms for manipulating the new structure. Perhaps an analog of the unate recursive paradigm can be developed in this more general setting. Many of the "special cases" which arise, for example, the "Achille's heel" function, are symptoms of our need for a data structure which reflects the true complexity (or simplicity) of the function being represented.

References

[AGA 80] V.K. Agarwal, "Multiple fault detection in programmable logic arrays," *IEEE Trans on Comp.*, Vol. C-29, No. 6, pp. 518-522, June 1980.

[ARE 78] Z. Arevalo and J. G. Bredeson, "A method to simplify a Boolean function into a near minimal sum-of-products for programmable logic arrays," *IEEE Trans. on Comp.*, Vol. C-27, No. 11, pp. 1028-1039, November 1978.

[ATK 83] D.E. Atkins, W. Liu and S. Ong, "Overview of an arithmetic design system," *Proc. 20th Design Automation Conference,* pp. 314-321, June 1983.

[AUG 78] M. Auguin, F. Boeri, and C. Andre, "An algorithm for designing multiple Boolean Functions: Application to PLA's.," *Digital Process,* No. 4, pp. 215-230, 1978.

[BAH 81] R. J. Bahnsen, "Essential prime implicant tester", *IBM Tech. Disclosure Bulletin,* Vol. 24, No. 5, pp. 2344, October 1981.

[BAR 77] M. Barbacci, D. Siewiorek, R. Gordon, R. Howbrigg and S. Zuckermann, "An Architectural Research: ISP Description, Simulation and Data Collection," *Proc. Nat. Comp. Conf.,* Vol. 46, 1977.

[BAR 81] M. Barbacci, "Instruction Set Processor Specifications (ISPS): The Notation and its Specification," *IEEE Trans. on Comp.,* Vol. C-30, pp. 24-40, January 1981.

[BES 81] P.W. Besslich and P. Pichlbauer "Fast transform procedure for the generation of near minimal covers of Boolean functions," *IEEE Proc.,* Vol. 128, Pt. E, No. 6, pp. 250-254, November 1981.

[BLU 79] R. Blumberg and S. Brenner, "A 1500-gate random-logic large scale integrated (LSI) masterslice," *IEEE Solid State Journal,* Vol. SC-14, pp. 818-822, October 1979.

[BOS 82] P. Bose and A. Abraham, "Test generation for programmable logic arrays," *Proc. 19th Design Automation Conference,* pp. 574-580, June 1982.

[BOS 83] A. Bose, B. Chawla and H. Gummel, "A VLSI design system," *Proc. Int. Symp. on Circ. and Syst.,* pp. 734-737, May 1983.

[BRA 82a] R. Brayton, G.D. Hachtel, L. Hemachandra, A.R. Newton and A.L. Sangiovanni-Vincentelli, "A comparison of logic minimization strategies using ESPRESSO. An APL program package for partitioned logic minimization," *Proc. Int. Symp. on Circ. and Syst.,* pp. 43-49, Rome, May 1982.

[BRA 82b] R. Brayton and C. McMullen, "The decomposition and factorization of Boolean expressions," *Proc. Int. Symp. on Circ. and Syst.,* pp. 49-54, Rome 1982.

[BRA 82c] R.K. Brayton, J.D. Cohen, G.D. Hachtel, B.M. Trager, and D.Y.Y. Yun, "Fast recursive Boolean function manipulation," *Proc. 1982 International Symposium on Circuit and Systems,* pp. 58-62, May 1982.

[BRA 84] R.K. Brayton, C. McMullen, to be published in *Proc. ICCD'84,* Rye N. Y., 1984.

[BRE 72] M.A. Breuer (Ed.), *Design Automation of Digital Systems,* Vol. 1: Theory and Techniques, Prentice-Hall, 1972.

[BRI 78] P. Bricaud and J. Campbell, "Multiple output PLA minimization: EMIN," WESCON 78, 33/3, 1978.

[BRO 81] D.W. Brown, "A State-Machine Synthesizer - SMS," *Proc. 18th Design Automation Conference,* pp. 301-304, Nashville, June 1981.

[BRO 83] M.W. Brown and M.J. Kimmel, "Multiple implementations of a microprocessor from a single high-level design," *International Conference on Computer Design,* pp. 674-677, 1983.

[CHA 78] C.W. Cha, "A testing strategy for PLA's," *Proc. 15th Design Automation Conference,* June 1978.

[CHU 82] S. Chuquillanqui and T. Perez Segovia, "PAOLA: A tool for topological optimization of large PLAs," *Proc. 19th Design Automation Conference,* pp. 300-306, Las Vegas, June 1982.

[CLA 75] C.R. Clare, "*Designing Logic Systems using State Machines,*" McGraw Hill, 1975.

[COO 79a] P.W. Cook, S.E. Shuster, J.T. Parrish, V. Di Lonardo and D.R. Freedman, "1 μm MOSFET VLSI technology: Part III - Logic circuit design methodology and applications," *IEEE Trans. on Elec. Dev.,* Vol. ED-26, No. 4, pp. 333-345, April 1979.

[COO 79b] P.W. Cook, C.W. Ho, and S.E. Schuster, "A study in the use of PLA-based macros," *IEEE J. Solid-State Circuits,* Vol. SC-14, No. 5, pp. 833-383, Oct. 1979.

[CUR 69] H.A. Curtis, "Systematic procedures for realizing synchronous sequential machines using flip-flop memory: Part 1," *IEEE Trans. on Comp.,* Vol. C-18, pp. 1121-1127, December 1969.

[CUR 70] H.A. Curtis, "Systematic procedures for realizing synchronous sequential machines using flip-flop memory: Part 2," *IEEE Trans. on Comp.,* Vol. C-19, pp. 66-73, January 1970.

[DAE 81] W. Daehn and J. Mucha, "A hardware approach to self-testing of large programmable logic arrays," *IEEE Trans. on Comp.,* Vol. C-30, pp. 829-833, 1981.

[DAR 81] J. Darringer, W. Joyner, C. Berman and L. Trevillyan, "Logic synthesis through local transformations," *IBM J. Res. and Dev.,* Vol. 4, pp. 272-280, 1981.

[DAV 83] M. Davio and others, *Digital Systems with Algorithm Implementation,* John Wiley and Sons, New York, 1983.

[DEG 83] A.C. De Graaf and R. Nouta, "Layout generation of PLA based circuits from a register transfer description," *International Conference on Computer Design,* pp. 519-522, 1983.

[DEM 83a] G. De Micheli and A. Sangiovanni-Vincentelli, "Multiple folding of programmable logic arrays," *Proc. Int. Symp. on Circ. and Syst.,* Newport Beach (CA), pp. 1026-1029, May 1983.

[DEM 83b] G. De Micheli and A. Sangiovanni-Vincentelli, "PLEASURE: A computer program for simple/multiple constrained/unconstrained folding of programmable logic arrays," *Proc. 20th Design Automation Conference,* Miami Beach (FL), pp. 530-537, June 1983.

[DEM 83c] G. De Micheli and A. Sangiovanni-Vincentelli, "Multiple constrained folding of programmable logic arrays: theory and applications," *IEEE Trans. on CAD of Int. Circ. and Syst.,* Vol. CAD-2, No. 3, pp. 167-180, July 1983.

[DEM 83d] G. De Micheli and M. Santomauro, "SMILE: A computer program for partitioning of programmed logic array," *Computer Aided Design,* No. 2, pp. 89-97, March 1983.

[DEM 83e] G. De Micheli and M. Santomauro, "Topological partitioning of programmable logic arrays," *Proc. Int. Conf. on Comp. Aid. Des.,* Santa Clara, pp. 182-184, September 1983.

[DEM 83f] G. De Micheli, A. Sangiovanni-Vincentelli and T. Villa, "Computer-aided synthesis of PLA-based finite state machines," *Proc. Int. Conf. on Comp. Aid. Des.,* Santa Clara, pp. 154-157, September 1983.

[DEM 84] G. DeMicheli, M. Hofmann, R. Newton, A. Sangiovanni-Vincentelli, "A system for the automatic synthesis of programmable logic arrays," in *Advances in Computer-Aided Engineering,* A. Sangiovanni-Vincentelli editor, Jay Press, 1984.

[DEU 83] J.T. Deutsch and A.R. Newton, "Data-flow based behavioral-level simulation and synthesis," *Proc. Int. Conf. on Comp. Aid. Des.*, pp. 63-64, Santa Clara, CA, September 1983.

[DIE 69] D.L. Dietmeyer and J.R. Duley, "Translation of a DDL digital system specification to Boolean equations," *IEEE Trans. on Comp.*, Vol. C-18, pp. 305-313, April 1969.

[DIE 78] D.L. Dietmeyer, "*Logic Design of Digital Systems (Second Edition),*" Allyn and Bacon Inc., Boston, 1978.

[DIE 80] D.L. Dietmeyer and M.H. Doshi, "Automated PLA synthesis of the combinatorial logic of a DDL description," *Journal of Design Automation and Fault-Tolerant Computing,* Vol 3, pp. 241-257, April 1980.

[DIE 79] D.L. Dietmeyer, "Connection arrays from equations," *Journal of Design Automation and Fault-Tolerant Computing,* Vol. 3, pp. 109-125, April 1979.

[DOL 64] T.A. Dolotta and E.G. McCluskey, "The coding of internal states of sequential machines," *IEEE Trans. Elect. Comp.*, Vol. EC-13, pp. 549-562, October 1964.

[EGA 82] J.R. Egan and C.L. Liu, "Optimal bipartite folding of PLA," *Proc. 19th Design Automation Conference,* pp. 141-146, Las Vegas, June 1982.

[EIC 77] E.B. Eichelberger and T.W. Williams, "A logic design structure for LSI testability," Proc. 14th Design Automation Conference, pp. 462-468, 1977.

[EIC 80] E.B. Eichelberger and E. Lindbloom, "A heuristic test-pattern generation for programmable logic arrays," *IBM J. Res. Develop.*, Vol. 24, No. 1, pp. 12-22, January 1980.

[EIC 83] E.B. Eichelberger and E. Lindbloom, "Random-pattern coverage enhancement and diagnosis for LSSD logic self-test," *IBM J. Res. and Dev.*, Vol. 27, No. 3, pp. 265-272, March 1983.

[ELL 82] S. Ellis, K. Keller, A. Newton, D. Pederson, A. Sangiovanni-Vincentelli, C. Sequin, "A symbolic layout design system," *Proc. Int. Symp. on Circ. and Syst.*, Rome, Italy, May 1982.

[FAN 83] S. Fang, *High Speed Bipolar PLA Design Techniques*, Ph.D. dissertation, U.C. Berkeley 1983.

[FEL 76] A. Feller, "Automatic layout of low-cost quick-turnaround random-logic custom LSI devices," *Proc. 13th Design Automation Conference*, pp. 79-85, June 1976.

[FLE 70] H. Fleisher, A. Weinberger and V. Winkler, "The writable personalized chip," *Computer Design*, Vol. 9, No. 6, pp. 59-66, June 1970.

[FLE 75] H. Fleisher and L.I. Maissel, "An introduction to array logic," *IBM J. Res. and Dev.*, Vol. 19, pp. 98-109, March 1975.

[FLE 79] H. Fleisher, L.I. Maissel and D.L. Ostapko, "A structured design methodology based on associative arrays," *Fourth International Symposium on CHDL*, pp. 89-95, October 1979.

[FLE 80] W. Fletcher, "*An Engineering Approach to Digital Design*," Prentice Hall, 1980.

[FLO 82] R. Floyd and J. Ullman, "The compilation of regular expressions into integrated circuits," *ACM Journal*, Vol. 29, No. 3, pp. 603-622, July 1982.

[FUJ 80] H. Fujiwara, K. Kinoshita and H. Ozaki, "Universal test sets for programmable logic arrays," *Digest 10th International Symposium on Fault-Tolerant Computing*, pp. 137-142, October 1980.

[FUJ 81] H. Fujiwara and K. Kinoshita, "A Design of programmable logic arrays with universal tests," Joint Special Issue on Design for Testability, *IEEE Trans. on Comp.*, Vol. C-30, No. 11, pp. 823-828; and *IEEE Trans. Circ. and Syst.*, Vol. CAS-28, No. 11, pp. 1027-1032, November 1981.

[GAR 79] M.R. Garey and D.S. Johnson, *Computers and Intractability*, W.H. Freeman and Company, San Francisco, 1979.

[GAU 77] J.W. Gault, "The generation of programmable logic array (PLA) implementation from higher level language description," *15th Annual Allerton Conference*, University of Illinois, pp. 120-125, 1977.

[GHA 56] M.J. Ghazala, "Irredundant disjunctive and conjunctive forms of a Boolean function," *IBM J. of Res. and Dev.*, Vol. 1, pp. 171-176, 1956.

[GIL 81] J.L. Gilkinson and S.D. Lewis, "Conversion of PLA-logic to random-logic implementations," *IBM Technical Disclosure Bulletin*, Vol. 24, No. 3, August 1981.

[GLA 80] L.A. Glasser and P. Penfield, Jr., "An interactive PLA generator as an archetype for a new VLSI design methodology," *Proceeding of International Conference on Circuits and Computers*, pp. 608-611, October 1980.

[GOL 80] R.L. Golden, P.A. Latus and P. Lowry, "Design automation and the programmable logic array macro," *IBM J. of Res. and Dev.*, Vol. 24, No. 1, pp. 23-31, January 1980.

[GRA 82] W. Grass, "A depth-first branch-and-bound algorithm for optimal PLA folding," *Proc. 19th Design Automation Conference*, pp. 133-140, Las Vegas 1982.

[GRE 76] D.L. Greer, "An associative logic matrix," *IEEE Journal of Solid State circuits*, Vol. SC-11, No. 5, pp. 679-691, October 1976.

[GRE 83] D.L. Greer, "Analysis of interconnection in traditional ICs and derivative PLA-like structures," *International Conference on Computer Design,* pp. 162-166, 1983.

[HAC 80] G.D. Hachtel, A.L. Sangiovanni-Vincentelli and A.R. Newton, "An algorithm for optimal PLA folding," *Proc. Int. Conf. on Circ. and Comp.,* pp. 1023-1028, New York, N.Y., October 1980.

[HAC 82] G.D. Hachtel, A.R. Newton, and A.L. Sangiovanni-Vincentelli, "An algorithm for optimal PLA folding," *IEEE Trans. on CAD of Int. Circ. and Syst.,* pp. 63-77, Vol. 1, No. 2, April 1982.

[HAL 78] A.K. Halder, "Grouping table for the minimization of n-variable Boolean functions," *Proc. IEEE,* Vol. 125, No. 6, pp. 474-482, June 1978.

[HAR 61] J. Hartmanis, "On the state assignment problem for sequential machines," *IRE Trans. Elect. Comp.,* Vol. EC-10, pp. 157-165, June 1961.

[HAR 66] J. Hartmanis and R.E. Stearns, "*Algebraic Structure Theory of Sequential Machines,*" Prentice Hall, 1966.

[HAS 82] S.Z. Hassan and E.J. McCluskey, "Testing PLA's using multiple parallel signature analyzers," *Digest 13th International Symposium on Fault-Tolerant Computing,* pp. 422-425, June 1982.

[HEN 83] J. Hennessy, "Partitioning programmable logic arrays. Summary," *Proc. Int. Conf. on Comp. Aid. Des.,* pp. 180-181, Santa Clara, CA, September 1983.

[HIL 78] F. Hill and G. Peterson, "*Digital Systems: Hardware Organization and Design,*" Wiley, 1978.

[HIL 81] F. Hill and G. Peterson, "*Introduction to Switching Theory and Logical Design,*" Wiley, 1981.

[HON 72] S.J. Hong and D.L. Ostapko, "On complementation of Boolean functions", *IEEE Trans. on Comp.,* Vol. C-21, pp. 1072, 1972.

[HON 74] S.J. Hong, R.G. Cain and D.L. Ostapko, "MINI: A heuristic approach for logic minimization," *IBM J. of Res. and Dev.*, Vol. 18, pp. 443-458, September 1974.

[HON 80] S.J. Hong and D.L. Ostapko, "FITPLA: A programmable logic array for functional independent testing," *Digest 10th International Symposium on Fault-Tolerant Computing,* pp. 131-136, October 1980.

[HU 83] T.C. Hu and Y.S. Kuor, "Graph folding and programmable logic array," *Computer Science Technical Report,* No. CS-71, UCSD 1983.

[ISH 82] K. Ishikawa, T. Sasao and H. Terada, "A minimization algorithm for logical expressions and its bounds of application," (in Japanese), *Trans. IECE Japan,* Vol. J65-D, No. 6, pp. 797-804, June 1982.

[ISH 83] K. Ishikawa, T. Sasao and H. Terada, "A simplification algorithm for logical expressions: A5," (in Japanese), *Trans. IECE Japan,* Vol. J 66-D, No. 1, pp. 41-48, January 1983.

[JON 75] J.M. Jones, "Array logic macros," *IBM J. of Res. and Dev.,* Vol. 19, pp. 120-126, March 1975.

[KAM 79] Y. Kambayashi, "Logic design of programmable logic arrays," *IEEE Trans. on Comp.*, Vol. C-28, No. 9, pp. 609-617, September 1979.

[KAN 81] Sungho Kang, "Automated synthesis of PLA based systems," *Ph.D. Dissertation,* Stanford University, 1981.

[KAN 81] S. Kang and W.M. vanCleemput, "Automatic PLA synthesis from a DDL-P description," *Proceedings of the 18th Design Automation Conference,* pp. 391-397, June 1981.

[KAN 83] S.M. Kang, R.H. Krambeck, H-F S. Law and A.D. Lopez, "Gate matrix layout of random control logic in a 32-bit CMOS CPU adaptable to evolving logic design," *IEEE Trans. on CAD of Int. Circ. and Syst.*, Vol. CAD-2, No. 1, pp. 18-29, January 1983.

[KAR 83] A.R. Karlin, H.W. Trickey and J.D. Ullman, "Experiance with a regular expression compiler," *International Conference on Computer Design,* pp. 656-665, 1983.

[KEL 82] K. Keller, A. Newton, "A symbolic design system for integrated circuits," *Proc. 19th Design Automation Conference,* June 1982.

[KEL 83] K. Keller, "An electronic circuit CAD framework," *Ph.D. Dissertation,* U.C. Berkeley 1983.

[KHA 82] J. Khakbaz, "A testable PLA design with low overhead and high fault coverage," *Digest 13th International Symposium on Fault-Tolerant Computing,* pp. 426-429, June 1982.

[KOV 79] T. Kovylarz and A. Al-Najjar, "An examination of the cost function for programmable logic arrays," *IEEE Trans. on Comp.,* Vol. C-28, No. 8, pp. 586-590, August 1979.

[LIP 83] H.M. Lipp, "Methodical aspects of logic synthesis," *Proceedings of IEEE,* Vol. 71, No. 1, January 1983.

[LOG 75] J.C. Logue, N.F. Brickman, F. Howley, J.W. Jones and W.W. Wu, "Hardware implementation of a small system in programmable logic arrays," *IBM J. of Res. and Dev.*, Vol. 19, pp. 110-119, March 1975.

[LUB 82] M. Luby, U. Vazirani, V. Vazirani and A. Sangiovanni-Vincentelli, "Some theoretical results on the optimal PLA folding problem," *Proc. Int. Conf. on Circ. and Comp.,* pp. 165-170, New York, N.Y., October 1982.

[MAH 83] G. Mah, "PANDA - A PLA generator for multiply folded arrays," *M.S. Report,* Dept. EECS, U.C. Berkeley 1983.

[MAI 82] L.I. Maissel and D.L. Ostapko, "Interactive design language: a unified approach to hardware simulation, synthesis and documentation," in *Proc. 19th Design Automation Conference,* 1982.

[MAI 83] L.I. Maissel and R.L. Phoenix, "IDL (Interactive Design Language) features and philosophy," *International Conference on Computer Design*, pp. 667-669, 1983.

[MCC 56] E.J. McCluskey, "Minimization of Boolean functions," *Bell Syst. Tech. Jour.*, Vol. 35, pp. 1417-1444, April 1956.

[MCC 65] E.J. McCluskey, Jr., *Introduction to the Theory of Switching Circuit*, McGraw-Hill, 1965.

[MCC 79] E.J. McCluskey, "Designing with PLA's," *Proceedings of 13th Asilomar Conference on Circuits, Systems, and Computers*, New York, pp. 442-445, 1979.

[MEA 80] C. Mead and L. Conway, "*Introduction to VLSI Systems,*" Addison Wesley, 1980.

[MEH 80] E.I. Mehldorf, G.P. Papp and T.W. Williams, "Efficient test pattern generation for embedded PLA's," *1980 IEEE Test Conference*, pp. 359-367, 1980.

[MIC 71] R.S. Michalski and Z. Kulpa, "A system of programs for the synthesis of combinatorial switching circuits using the method of disjoint stars," *Proceedings of International Federation of Information Processing Societies Congress*, 1971, Booklet TA-2, p. 158, 1971.

[MOR 70] E. Morreale, "Recursive operators for prime implicant and irredundant normal form determination," *IEEE Trans. on Comp.*, Vol. C-19, p. 504, 1970.

[MOR 83] M. Morisue, K. Isaji, H. Fukuzawa and Z.H. Shaikh, "A novel approach for designing PLA," *International Conference on Computer Design*, pp. 566-569, 1983.

[MUR 79] S. Muroga, *Logic Design and Switching Theory*, Wiley-Interscience Publication, 1979.

[MUR 82] S. Muroga *VLSI System Design*, John Wiley and Sons, New York, 1982.

[NAG 75] H.T. Nagale, Jr., B.D. Carrol, and J.D. Irwin, *An Introduction to Computer Logic,* p. 168, Prentice-Hall, Inc., 1975.

[NAK 78] S. Nakamura, S. Murai and C. Tanaka, "LORES-logic reorganization system," in *Proc. 15th Design Automation Conference,* pp. 250-260, 1978.

[NEW 81] A.R. Newton, D.O. Pederson, A.L. Sangiovanni-Vincentelli and C.H. Sequin, "Design aids for VLSI: the Berkeley perspective," *IEEE Trans. on Circ. and Syst.,* Vol. CAS-28, pp. 618-633, July 1981.

[OST 74] D.L. Ostapko and S.J. Hong, "Generating test examples for heuristic Boolean minimization," *IBM J. of Res. and Dev.,* Vol. 18, pp. 459-464, September 1974.

[OST 79] D.L. Ostapko and S.J. Hong, "Fault analysis and test generation for programmable logic arrays," *IEEE Trans. on Comp.,* Vol. C-28, No. 9, pp. 617-626, September 1979.

[PAI 81] J.F. Paillotin, "Optimization of the PLA area," *Proc. 18th Design Automation Conference,* pp. 406-410, Nashville, June 1981.

[PAP 83] C.A. Papachristou and D. Sarma "An approach to sequential circuit construction in LSI programmable arrays," *IEEE Proc.,* Vol. 130, Pt. E, No. 5, pp. 159-164, September 1983.

[PAT 79] S. Patil and T. Welch, "A programmable logic approach for VLSI," *IEEE Trans. on Comp.,* Vol. C-28, No. 9, pp. 594-601, September 1979.

[PAT 83] S. Patil, C. Leung, B. Knapp and H. Ravindra, "Strage/Logic Array (SLA): A high density standard cell design method," *International Conference on Computer Design,* pp. 167-171, 1983.

[PER 76] G. Persky, D.N. Deutsch and D.G. Schweikert, "LTX - A system for the directed automation design of LSI circuits," in *Proc. 13th Design Automation Conference*, 1976.

[PER 77] G. Persky, D.N. Deutsch and D.G. Schweikert, "LTX - A minicomputer-based system for automatic LSI layout," *J. Design Automation and Fault-Tolerant Computing*, Vol. 1, No. 3, pp. 217-255, May 1977.

[POM 79] G. Pomper and R.J. Armstrong, "An efficient multivalued minimization algorithm," *Proceedings ISMVL-79*, May 1979.

[RAM 83] K.S. Ramanatha and N.N. Biswas, "A design for testability of undetectable crosspoint faults in programmable logic arrays," *IEEE Trans. on Comp.*, Vol. C-32, No. 6, pp. 551-557, June 1983.

[REN 75] B. Rensch, "Generation of prime implicants from subfunctions and a unifying approach to the covering problem," *IEEE Trans. on Comp.*, Vol. C-24, No. 9, pp. 926-930, September 1975.

[RHY 77] V.T. Rhyne, P.S. Noe, M.N. McKinny, and U.W. Pooch, "A new technique for the fast minimization of switching function," *IEEE Trans. on Comp.*, Vol. C-26, No. 8, pp. 757-764, August 1977.

[ROT 58] J.P. Roth, "Algebraic topological methods for the synthesis of switching functions," *Trans. Amer. Math. Soc.*, Vol. 88, pp. 301-326, July 1958.

[ROT 59] J.P. Roth, "Algebraic topological methods in synthesis," in *Proceeding of International Symposium on Theory of Switching*, April 1957. In *Annals of Computational Laboratory of Harvard University*, Vol. 29, pp. 57-73, 1959.

[ROT 78] J.P. Roth, "Programmed logic array optimization," *IEEE Trans. on Comp.*, pp. 174-176, February 1978.

[ROT 80] J.P. Roth, "*Computer Logic, Testing and Verification*," Computer Science Press, 1980.

[SAL 82] K.K. Saluja, K. Kinoshita and H. Fujiwara, "A multiple fault testable design of programmable logic arrays," *Digest 10th International Symposium on Fault-Tolerant Computing*, pp. 44-46, June 1982.

[SAS 78] T. Sasao, "An application of multiple-valued logic to a synthesis of programmable logic arrays," *Proceedings of ISMVL-78*, pp. 65-72, May 1978.

[SAS 79] T. Sasao and H. Terada, "Multiple-valued logic and the design of programmable logic arrays with decoders," *Proceedings of ISMVL-79*, pp. 27-37, May 1979.

[SAS 80] T. Sasao and H. Terada, "On the complexity of shallow logic functions and the estimation of programmable logic array size," *Proceedings of ISMVL-80*, pp. 65-73, June 1980.

[SAS 81] T. Sasao, "Multiple-valued decomposition of generalized Boolean functions and the complexity of programmable logic arrays," *IEEE Trans. on Comp.*, Vol. C-30, No. 9, pp. 635-643, September 1981.

[SAS 82] T. Sasao, "An application of multiple-valued logic to a design of masterslice gate array LSI," *Proc. of 12th International Symposium on Multiple-Valued Logic*, pp. 45-54, May 1982.

[SAS 83a] T. Sasao, "A fast complementation algorithm for sum-of-products expressions of multiple-valued input binary functions," *Proc. of 13th International Symposium of Multiple-Valued Logic*, pp. 103-110, May 1983.

[SAS 83b] T. Sasao, "Input variable assignment and output phase optimization of PLA's," Mathematical Science Department, IBM Thomas J. Watson Research Center, RC 1003, June 1983.

[SAS 83c] T. Sasao, "A hardware for logic minimization," (in Japanese), *The 9th Workshop on FTC*, July 1983.

[SAS 83d] T. Sasao, Private Communication, 1983.

[SAS 84] T. Sasao, S.J. Hong, and R.K. Brayton, "Minimization of PLA's by decomposition," (in preparation).

[SAU 72] G. Saucier, "State assignment of asynchronous sequential machines using graph techniques," *IEEE Trans. on Comp.,* Vol. C-21, pp. 282-288, March 1972.

[SCH 79] M.S. Schmookler, "A scheme for implementing microprogram addressing with programmable logic arrays," *Digital Process,* No. 5, pp. 235-256, 1979.

[SCH 80] M.S. Schmookler, "Design of large ALUs using multiple PLA macros," *IBM J. of Res. and Dev.*, Vol. 24, pp. 2-14, January 1980.

[SHA 48] C. E. Shannon, "The synthesis of two-terminal switching circuits", *Bell Sys. Tech. J.*, 1948.

[SHE 77] W. Sherwood, "PLATO-PLA Translator/Optimizer," *Proceedings of Design Automation and Microcomputers,* pp. 28-35, 1977.

[SIG 79] *Signetics Bipolar and MOS Memory Data Manual,* pp. 156-188, 1979.

[SMI 79] J.E. Smith, "Detection of faults in programmable logic arrays," *IEEE Trans. on Comp.,* Vol. C-28, No. 11, pp. 845-853, November 1979.

[SMI 83] K.F. Smith, "Design of regular arrays using CMOS in PPL," *International Conference on Computer Design,* pp. 158-161, 1983.

[SOM 82] F. Somenzi, S. Gai, M Mezzalama and P. Prinetto, "PART: Programmable array testing based on a PARTitioning algorithm," *Digest 13th International Symposium on Fault-Tolerant Computing,* pp. 430-433, June 1982.

[SON 80] K. Son and K.K. Pradhan, "Design of programmable logic arrays for testability," *1980 IEEE Test Conference,* pp. 163-166, 1980.

[STI 83] D.W. Still and R.C. Weintritt, "An automatic LASER personalization system for PLAs," *International Conference on Computer Design,* pp. 101-104, 1983.

[SU 72] S.Y.H. Su and P.T. Cheung, "Computer minimization of multi-valued switching functions," *IEEE Trans. on Comp.,* Vol. 21, pp. 995-1003, 1972.

[SUW 81] I. Suwa and W.J. Kubitz, "A computer aided design system for segment-folded PLA macro cells," *Proc. 18th Design Automation Conference,* pp. 398-405, Nashville, June 1981.

[SVO 79] A. Svoboda and D.E. White, "*Advanced Logical Circuit Design Techniques,*" Garland Press, New York, 1979.

[TIS 67] P. Tison, "Generalization of consensus theory and application to the minimization of Boolean functions," *IEEE Trans. on Elect. Comp.,* Vol. EC-16, pp. 446-456, 1967.

[VER 80] B. Vergineres, "Macro generation algorithm for LSI custom chip design," *IBM J. of Res. and Dev.,* Vol. 24, No. 5, pp. 612-621, September 1980.

[WAN 82] S.L. Want and A. Avizienis, "The design of totally self-checking circuits using programmable logic arrays," *Digest 9th International Symposium on Fault-Tolerant Computing,* pp. 173-180, June 1982.

[WEB 79] H. Weber, "High level design of programmed logic arrays," *Proc. Int. Symp. on CHDL,* October 1979.

[WEI 67a] A. Weinberger, "Large scale integration of MOS complex logic: a layout method," *IEEE Sol. St. Jour.,* Vol. SC-2, No. 4, pp. 182-190, December 1967.

[WEI 79] A. Weinberger, "High-speed programmable logic array adders," *IBM J. of Res. and Dev.,* Vol. 23, pp. 163-178, March 1979.

[WEI 67b] P. Weiner and E.J. Smith, "Optimization of reduced dependencies for synchronous sequential machines," *IEEE Trans. on Elect. Comp.*, Vol. EC-16, pp. 835-847, December 1967.

[WOO 75] R.A. Wood, "High-speed dynamic programmable logic array chip," *IBM J. of Res. and Dev.*, Vol. 19, pp. 379-383, July 1975.

[WOO 79] R.A. Wood, "A high density programmable logic array chip," *IEEE Trans. on Comp.*, Vol. C-28, pp. 602-608, September 1979.

[WOO 83] K.R. Woodruff and P.S. Balasubramanian, "Top-down design using IDL," *International Conference on Computer Design*, pp. 670-673, 1983.

[YAJ 82] S. Yajima and T. Aramaki, "Autonomously testable programmable logic arrays," *Digest 10th International Symposium on Fault-Tolerant Computing*, pp. 41-43, June 1982.

Index

Procedures

ABOUT THE AUTHORS

Robert K. Brayton obtained his Ph.D. in Mathematics from MIT in 1961. He has been with the Mathematical Sciences Department of the IBM T. J. Watson Research Center since that time. In 1966 he was Visiting Professor at MIT and in 1975 at Imperial College, London. He is a Fellow of IEEE. Dr. Brayton received an IEEE Circuits and Systems Society Best Paper Award in 1971 and several Outstanding Invention Awards from IBM for his work on computer-aided design. He coauthored a book on systems and optimization in CAD and is the coauthor of over 60 technical papers.

Gary D. Hachtel received his Ph.D. from the University of California at Berkeley in 1964. He was with the IBM T. J. Watson Research Center from 1964 to 1981. In 1981 he joined the University of Colorado at Boulder where he is Professor of Electrical Engineering and Computer Sciences. He received an IEEE Circuits and Systems Society Best Paper Award in 1971 and the IEEE Baker Award for the Best Paper published on IEEE transactions and proceedings. He is a Fellow of the IEEE. He has written more than 50 papers in the area of computer-aided design.

Curtis T. McMullen is a graduate student in mathematics at Harvard University. He spent a year at Cambridge University, England, under a Herchel Smith Fellowship. He has worked on computer-aided design problems for the past five years in collaboration with Dr. Robert K. Brayton. His research interests, aside from the combinatorics of VLSI design, include ergodic theory and conformal dynamics.

Alberto L. Sangiovanni-Vincentelli received his "Dottore in Ingegneria" degree from the Politecinco di Milano, Italy, summa cum laude in 1971. He was with the Politecino di Milano from 1971 to 1976. In 1976 he joined the Department of Electrical Engineering and Computer Sciences of the University of California, Berkeley, where he is a Professor. In 1980 he was Visiting Scientist at the IBM T. J. Watson Research Center. He is an IEEE Fellow and has received a number of awards, including the Distinguished Teaching Award from the University of California in 1981, the Corillemin Lerner Award for the best paper published in IEEE transactions sponsored by the Circuits and Systems Society in 1982, two best paper awards and a best presentation award at the Design Automation Conference in 1982 and 1983. He has written over 90 papers in CAD.